GW01464437

A HISTORY OF THE NORTH WALES SLATE INDUSTRY

A HISTORY OF THE NORTH WALES SLATE INDUSTRY

Jean Lindsay

DAVID & CHARLES
NEWTON ABBOT LONDON
NORTH POMFRET (VT) VANCOUVER

ISBN 0 7153 6264 X

For

DAVID

Set in 11 on 13 point Imprint and printed in Great Britain by Latimer Trend & Company Ltd Plymouth for David & Charles (Holdings) Limited South Devon House Newton Abbot Devon

Published in the United States of America by David & Charles Inc North Pomfret Vermont 05053 USA

Published in Canada by Douglas David & Charles Limited 3645 McKechnie Drive West Vancouver BC

CONTENTS

ILLUSTRATIONS

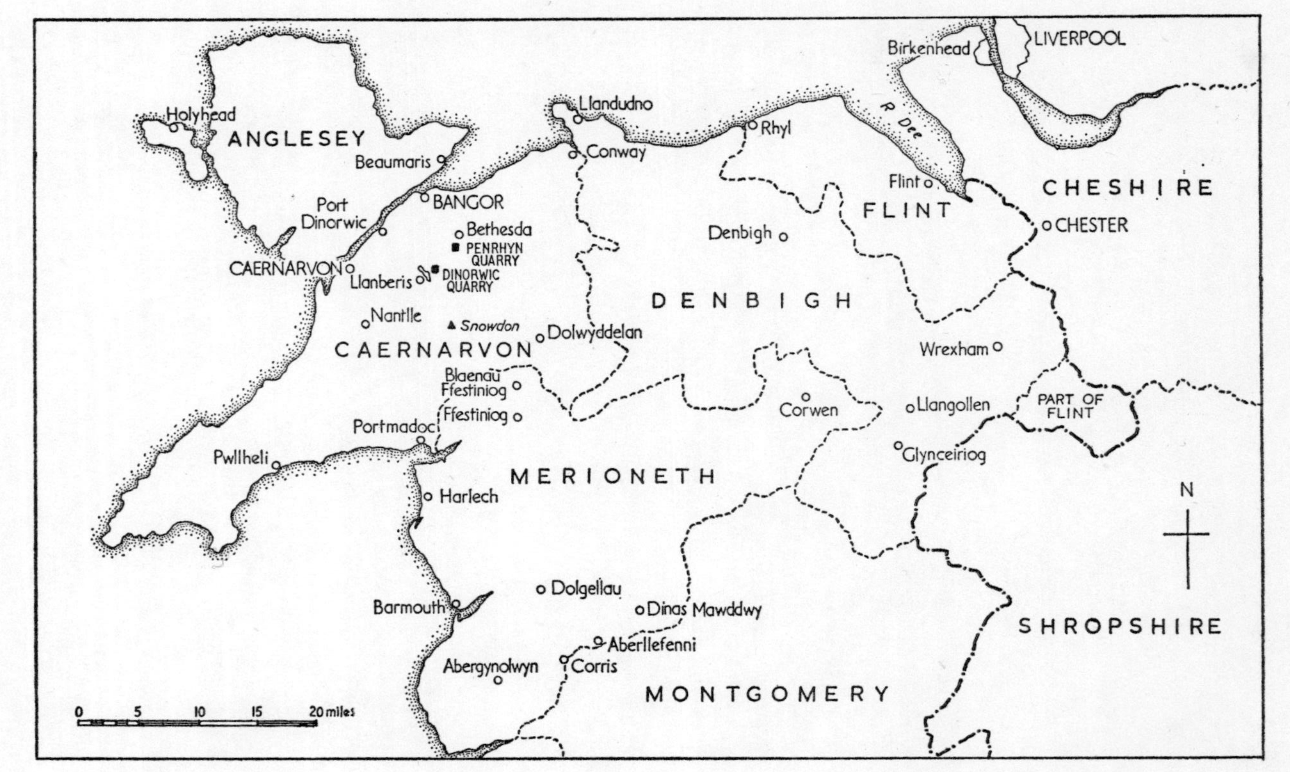

LIVERPOOL
Birkenhead
Holyhead
ANGLESEY
Llandudno
Rhyl
R Dee
Beaumaris
Conway
Flint
CHESHIRE
BANGOR
Port
Dinorwic
FLINT
CHESTER
Bethesda
PENRHYN QUARRY
Denbigh
CAERNARVON
Llanberis
DINORWIC QUARRY
DENBIGH
Nantlle
Snowdon
Dolwyddelan
Wrexham
CAERNARVON
Blaenau Ffestiniog
Corwen
Llangollen
PART OF FLINT
Ffestiniog
Portmadoc
Pwllheli
Glynceiriog
MERIONETH
Harlech
N
Dolgellau
Barmouth
Dinas Mawddwy
SHROPSHIRE
Aberllefenni
Abergynolwyn
Corris
MONTGOMERY
0
5
10
15
20 miles

I
THE SLATE INDUSTRY TO 1731

The development of the slate industry in North Wales was in many ways a remarkable achievement. The location of most of the important regions of slate was in Snowdonia, a mountainous area containing fourteen peaks over 3,000ft, with Snowdon attaining 3,560ft. This, even in the days of modern methods of transportation, created serious problems and meant also that the workmen had to face great change and variety of weather.

Before 1750 North Wales was a remote and wild country, with roads in the mountainous areas so poor that sledges, without wheels, were used to carry peat, slates and other minerals in Merionethshire and in parts of Caernarvonshire; and even in 1798 these sledges were noted by travellers. The Rev Richard Warner, on his second walk through Wales, described them as having 'the shape similar to the body of a waggon, capable of containing two or three hundredweight of peat'. He saw them in operation at Bwlch y Groes, where they were being used to collect peat from the mountains, and were drawn by 'a little stout Welsh pony'. Every cottager, Warner said, almost without exception, owned a pony; and he added that until the turnpike road was made, 'the use of wheels was scarcely known here, the sled being the only vehicle made use of'.[1]

Roads throughout Great Britain in the early eighteenth century were a general subject for complaint, but North Wales in many cases, had no roads at all. Another clergyman, the Rev W. Bingley, in his tour round North Wales in the summer of 1798, recorded that there was no carriage road from Caernarvon nearer to Llanberis than the bottom of the lower lake, 'the road from thence being nothing more than a horse path, and one of the worst I ever saw'.[2] For the first half of the century roads in North Wales remained under the control of the parishes, and statute labour and untrained surveyors continued to be employed. This meant that trade was prevented, and farmers and industrialists had only a local market, or a limited outlet to Ireland, or a coastwise trade to Chester and Liverpool, from the ports chiefly of Caernarvon and Beaumaris. After 1750, however, improvements were made, and new roads were constructed by private individuals and by public subscription, culminating in the formation of turnpike trusts, the first in North Wales probably being between Shrewsbury and Wrexham in 1752.[3]

Geographical obstacles also helped to isolate North Wales from England and from South Wales. The marshes of the Dee and the Severn hindered communication with England; and the central moorland plateau and the estuaries penetrating far inland hindered trade with the south. The uplands of Caernarvonshire and Merionethshire were suitable only for sheep pasture, and the resources were mainly scenery (which by the end of the century had attracted numerous travellers), water supplies, granite and slate. It is not, therefore, surprising that the slate industry did not expand in what is traditionally regarded as the beginning of the Industrial Revolution, that is 1760 to 1800; instead, the main period of expansion was from about 1831 to 1882. One is surprised only at the comparative speed with which the industry achieved its success, and at its almost equally rapid decline in the twentieth century.

The resources of Wales had been described as early as 1387 by John Trevisa in his translation of Ranulf Higden's *Polychronicon*:

Valeys bryngeth forth food,
And hills metal right good,
Col groweth under lond,
And grass above at the hond,
There lyme is copious,
And sclattes also for hous.[4]

The Welsh standard of living, however, compared with much of England, was low. Houses, in general, were primitive, and were often shared with the animals, and Warner, in 1798, found that in Merionethshire the pig was 'generally considered as *one* of the *family*' and was 'very commonly seen reposing comfortably before the cottage fire, with the children sporting round him'.[5] J. Evans, in his tour through North Wales at the end of the same century, described the houses in Bettws Garmon and in the Waunfawr district as having no doors, and said that 'an aperture in the roof' served for a chimney.[6] Thomas Pennant, in 1810, noted that the houses of the 'common people' in Lleyn were 'very mean; made with clay, thatched, and destitute of chimnies'.[7] The Welsh diet was also poor: even in 1839, Leigh's *Guide* informed would-be travellers that the Welsh lived 'very abstemiously; potatoes and sour milk, bacon and oat-porridge; a dark heavy bread, or cake, with various admixtures of vegetable and farinaceous substances' composed the usual food of the people. Leigh's *Guide* mentioned the famous Welsh ale, called 'cwrw' made from barley.[8] This was described in 1795 by J. Hucks, another traveller, as 'a heavy glutinous ale', having 'charms enough to debauch the senses of the whole principality'.[9]

By comparison with the Welsh coal, iron, copper and lead industries, the slate industry was very backward. These other industries attracted capitalists in the second part of the eighteenth century, largely because of the availability of roads, canals and mineral railways. The coal industry, by the end of the eighteenth century, had established itself at Wrexham, Holywell, Mold and Ruabon; and the iron industry, mainly under John Wilkinson's organisation, had been created as a large-scale concern at Bersham and at neighbouring Brymbo. The Welsh copper industry, largely as a result of the impetus given by the demands of the navy for copper sheathing in 1761, rose to great heights under the leadership of Thomas Williams. The opening up of the Parys Mountain in Anglesey by the Mona Company led to the growth and prosperity of the port of Amlwch; and by the closing years of the eighteenth century, Holywell was carrying on all branches of copper manufacture.[10] The lead industry had flourished most successfully in the seventeenth century in Caernarvonshire, when local ores were shipped from Conway. After 1850 the industry was again prosperous. In Flintshire lead mines were producing large quantities in the second half of the eighteenth century, and much of this lead and lead ore was being exported from Chester.[11]

In the slate industry, by contrast, development proceeded slowly until just before the start of Queen Victoria's reign; but the annual output of slates in North Wales rose from about 100,000 tons in 1832 to over 450,000 tons in 1882. This, by 1972, had fallen to about 22,000 tons; and at present what was once known as 'common slate', ie Welsh slate, is becoming increasingly uncommon in building. Substitutes, mainly tiles and concrete, have almost ousted what was once a universal roofing material.

The word 'slate' is derived from late Middle English 'slat' or 'sclate'; but after about 1630 the form with scl or skl became exclusively northern or Scottish. The word is connected with the French verb 'esclater', to break into pieces, referring to the characteristic property of splitting into thin sheets.[12] All fissile rocks, including slate, can be called 'llech' in Welsh; but the older Welsh name 'ysglatus, ysglats, or sglatys' was used only for slates, and dates certainly from the time of Guto'r Glyn. This fifteenth-century poet wrote a poem asking the Dean of Bangor, Dr Richard Kyffin (1480 to 1502), to send him a shipload of 'sglatys' from Aberogwen (near Bangor) to Rhuddlan, to roof a house at Henllan near Denbigh. In the translation of this poem by Dr Iorwerth Peate the slates are represented picturesquely as 'jewels from the hillside' and as 'warm slabs, as a crust on the timber of my house'.[13]

The geological definition of slate rests on the type of 'cleavage' (in Welsh 'hollt') which is known as 'slaty cleavage' to distinguish it particularly from that shown in many crystalline minerals. This means that the slate can be split into strong even sheets of almost any size, in a direction unrelated to the bedding planes. In Wales the counties of Caernarvon, Denbigh, Merioneth, Montgomery and Pembroke contain slate of economic value; and in other parts of Great Britain, the chief slate-producing areas are in Cornwall, Devon, the Lake District, Aberdeenshire and Argyll.

All the Welsh slate was formed during the Palaeozoic Era, in the time when most of what is now called North Wales was buried beneath the sea. During the Cambrian Period, layers of mud, sand and pebbles collected to a depth of about 5,000ft in Caernarvonshire and around 13,000ft in Merionethshire. Subsidence occurred on the sea floor during the Ordovician Period; and the fine-grained deposits were covered by lava from volcanoes which erupted on the bed of the sea. These processes of sinking and deposition resulted in thick accumulations of sediment helping to fill in the earth's crust, thus producing depressions or geosynclines. In North Wales the accumulation of sediment continued, until what was later to become slate was buried to a depth of about 25,000ft and because of the great weight and the heat was subjected to massive pressure.

At the end of the Silurian Period and during the succeeding Devonian Period, this part of the earth's crust was subjected to violent disturbances, the rocks already formed being pressed together; these earth-movements, known as the Caledonian, continued for some time. The sedimentary and volcanic rocks formed during the Cambrian, Ordovician and Silurian Periods were folded by the pressure into mountains, and these formed part of a chain extending in an approximately north-east to south-west direction. The Padarn ridge, from Bethesda to Penygroes, constituted a hard mass, and the softer sediments were squeezed against it, leading to their conversion into slate; as a result, in this region, the best slates were found where the Cambrian deposits were finest in grain and where they were closest to the ridge they had been pressed against.[14]

The fold caused by the Caledonian earth-movements, in the strata lying between the pre-Cambrian rocks of the Padarn ridge and the Cambrian grits of the Harlech dome, was called the Snowdonian syncline. The Harlech dome stretches from Ffestiniog in the north to the Mawddach estuary in the south, and from the coast on the west to Arenig Fawr and Rhobell Fawr in the east. To the east of this dome was another syncline, with its axis passing through Central Wales. The Bethesda-Llanberis Nantlle slate belt was found on the north-western side of the Snowdonian syncline; and the Blaenau Ffestiniog belt along the south-eastern; the Towyn–Dinas Mawddwy belt occurred on one of the sides of the Central Wales syncline.

Thus, the five main areas where slate of commercial importance is found are: firstly, in central Caernarvonshire, including Bethesda, Llanberis and Nantlle where the main slate beds are of the Cambrian Age; secondly, the area between Blaenau Ffestiniog and Capel Curig in Merionethshire and the adjacent parts of Caernarvonshire where the slates are of the Ordovician Age; thirdly, the country between Towyn and Dinas Mawddwy, around Corris in southern Merionethshire and the adjacent parts of Montgomeryshire where the Ordovician and Silurian strata yield slates; fourthly, the country between Llangollen and Corwen in Denbighshire and eastern Merionethshire where the slates are of the Silurian Age; and fifthly, the Presely district of Pembrokeshire and the neighbouring parts of Carmarthenshire, where the slates are of the Ordovician Age.

The slate beds of Caernarvonshire in the area stretching from Bethesda via Llanberis, Moel Tryfan and Nantlle to Penygroes are of the Cambrian formation, and they include those of the two largest slate quarries in the world, namely Penrhyn and Dinorwic, which are on opposite sides of the mountain Elidir Fawr. Dinorwic is on the south-west overlooking Llyn Peris and Penrhyn is on the north-east overlooking the Ogwen Valley. The Glynrhonwy Quarries are on the opposite side of the valley to Dinorwic above Llyn Padarn. In the Nantlle Valley, about six miles south-west of Llanberis, the slate outcrops on the floor of the valley, and the quarries are worked not in open terraces, as at Penrhyn and Dinorwic, but in deep pits. Dorothea and Penyrorsedd Quarries are located in the Nantlle Valley.

The most important slates of the Ordovician rocks of Snowdonia occur around Blaenau Ffestiniog; and the slate formation in most cases in Merionethshire dips deep down into the earth under the mountains, so that the slate is mostly mined. To the north-east of Blaenau Ffestiniog are the old quarries at Penmachno and at Dolwyddelan. Other commercial slates of the Ordovician Age occur in the region from Criccieth and Tremadoc in the south-west to the River Conway near Trefriw in the north-east. In the past roofing slates from the shales in the Ordovician System in Anglesey have been quarried at Llaneilian near Amlwch.[15]

The Pleistocene Ice Age considerably altered the topography of North Wales, and incidentally added to the geographical difficulties of the slate industry. Huge masses of ice moved south over the Irish Sea, and a Welsh ice sheet descending from the mountains cleared out valleys. During the latter part of the Ice Age local glaciers descended to form impressive valleys, lakes and moraines.[16]

There is a great deal of variation in the colour of the slates, even in an individual quarry. The slates can be divided into the following colours in the Llanberis area: 'green vein; silky red vein; hard spotted blue vein; royal blue vein; curly red vein; old blue vein; red hard vein; and Glynrhonwy vein'. In the Nantlle area the colours are as follows: 'green vein; silky vein; blue mottled vein; red and blue striped vein; red spotted vein; and red vein'. In the Penrhyn area the colours are: 'green; purplish blue; grey; grey-mottled; mottled and striped; and blue'. In the Ffestiniog area the colour is mainly blue-grey. In Corris, Abergynolwyn and Aberllefenni the colour is pale blue; and blue slate is found in Glyn Ceiriog and near Llangollen.[17]

The slate industry did not exist as an organised concern until the second half of the eighteenth century, but there is evidence that slate has been used for roofing since the Roman occupation. The rock has, from time to time, been put to many other uses: domestic implements made of slate dating from the early Iron Age and the Roman period have been discovered in North Wales. In 1944, for instance, a slate spindlewhorl, ie a small perforated disc used to increase the moment of inertia of the spindle used in hand spinning, was found at Caernarvon; it is now preserved in the National Museum of Wales.[18] It is probable also that slates were used as bakestones from medieval times. Slate, according to Gruffydd Ellis, a quarry agent, was obtained from

Dinorwic Quarry by the quarrymen for this purpose. It was used to bake bread about an inch thick known as 'bara llech' or 'slate bread'; and this practice continued until about 1780, when iron bakestones were introduced.[19]

It is generally accepted that the Romans were the first to use the purple slates of the Cambrian formation in Caernarvonshire. R. E. Mortimer Wheeler has given a description of the Roman fort of Segontium, just outside Caernarvon; and Dr E. Greenly, who assisted in the process of identification, noted that the Romans had not used slate merely because it was on the site, since the nearest outcrops of slate were five miles from Segontium. They evidently appreciated the excellent qualities of this slate when they used it for flooring and roofing. The later levels of the fort contained numerous slates, but no tiles, whereas the earliest contained roof-tiles, but no slates. This indicated that the local clays were worked by the Romans from the earliest times, and that slates were a later discovery.

In the third century at Segontium there was probably a change to slate: the cellar in the sacellum was floored with slate slabs, and the channels of an adjacent hypocaust were lined with slate. In the period of rebuilding in the fourth century slates were used to line the pit in one of the commandant's rooms, and broken slates mixed with other materials were used as metalling for the courtyards of the commandant's house and for the praetorium. Earlier than the slates at Segontium were the slaty flagstones used at Caer Llugwy, a Roman fort between Capel Curig and the Vale of Conway; but they were of a different type, belonging not to the Cambrian System but to the Silurian System. Occupation of this fort probably ended before or early in the second century AD.[20]

One of the oldest slate quarries in Wales is reputed to be the Cilgwyn Quarry in Nantlle. This is believed to have had its origin in the twelfth century; and when Edward I visited the copper mines of Drws y Coed in Nantlle Valley, he stayed at a house in Nantlle which was roofed with slates from the quarry. It is probable that these early slates were rough and thick having been obtained where the slate rock outcropped.[21]

Slate did not become a universal roofing material in Wales until the nineteenth century. Before that date most small dwellings, and even some churches, had thatched roofs. Edmund Hyde Hall, in his account of Caernarvonshire (1809–1811), described the church of Llanbedr as 'now slated, but . . . thatched within the time of living memory'; and the church in Penrhos as being 'wretchedly thatched . . . in no wise distinguishable from the cottages scattered about'.[22]

Before 1750 traditional roofing materials depended largely on the local materials available; and since slate was a heavy material to transport, lack of cheap carriage made it proportionately very expensive. The making of clay tiles came to an end after the Romans withdrew from Britain; but in the thirteenth century they were being manufactured again. In 1212 the Great Fire of London raged for twelve days, destroying a considerable part of the city and part of the town of Southwark. On 24 July a meeting was held in the Guildhall; and it was declared that 'Every person who should build a house' should take care that he did not cover it with reeds, rushes, stubble or straw, but only with tiles, shingles, board or lead.[23]

Caernarvon Castle, which, unlike Conway, was built slowly, was begun about 1284; and most of the towers, such as the Eagle Tower, the Blake Tower, the Banner Tower and the Gatehouse, had lead roofs. Around February 1317, however, the hall of Llewelyn, the last native Prince, was removed to Caernarvon from Conway and erected in the castle; and its roof was there slated by 'Henry le Sclatiere'. From 1322 Edward II's political troubles became so numerous that the castles were neglected; and by the middle of the fourteenth century the building period of Caernarvon Castle was over. In 1609 Sir John Haryngton reported that the castle was of no use for anything but a gaol.[24]

Conway Castle was begun in 1283 and most remarkably was substantially completed by 1287. It seems that as originally built all the internal buildings, and probably the eight towers of the castle, were roofed with shingles or slate. Purple slate was used in places for bedding and levelling courses in the thirteenth century work, and perhaps for much of the original roofing. This slate probably came from Aberogwen by boat, or by cart from Llangelynin three miles away. In 1525, in Henry VIII's time, the tower roofs were renewed; and at that date 5s was paid for 1,000 slates, with their carriage by boat from Aberogwen, 'to sclate the oder side of the well in the castell'. Lead had been substituted for timber throughout the castle in 1346; but in 1665, when the castle was in a state of decay, the lead and timber were sold, and the building was partly dismantled. As the castle was thenceforward roofless, there is now no clear evidence as to the original form of the roofs of the towers.[25]

In the fourteenth century, around 1358–60, building repairs were carried on in Chester under the supervision of the Black Prince; and it is recorded that '21,000 slate stones' were acquired for the roof of the great stable in Chester Castle. These stones were purchased at Ogwen at a cost of 38s 6d; and they were brought by water to Shotwick and thence by land to Chester, the total cost of transport being £2 7s 3d. The lower part of the Dee was important for the transport of large quantities of slate from the quarries of Ogwen to Chester.[26] As the Dee silted up, Chester's importance as a port declined; and in 1798 its trade in slates, copper, lead, calamine and potatoes was said to be inconsiderable.[27]

Apart from these castles, slate was not commonly used as a building material in Wales before the sixteenth century. An interesting exception to this, however, was the Bishop's Palace of Gogarth near Llandudno, which was a manor or grange belonging to the Bishops of Bangor. For the original construction, in the late thirteenth century, fairly thin purple roofing slate of the kind found at Bethesda was used; and some thin pieces were also used for bedding the dressed stone of the door.[28]

Another exception to this generalisation was the town of Conway, which contained houses with slate roofs from an early date. In a metrical chronicle for the year 1399 written by a Frenchman named Creton, Richard II's arrival at Conway was thus depicted:

> So rode the King, without making noise,
> That at Conway, where there is much slate

On the houses, he arrived, with scarce a pause,
At break of day.[29]

It is probable that this slate had been supplied by the quarries at Penmachno or Dolwyddelan. These quarries were still sending slate to Conway in the nineteenth century; but the slates had to be brought by carts to Trefriw, taken to Conway in sailing boats and then put into larger vessels. The local industry therefore declined, as it could not compete with quarries nearer to ports; and in 1876 the output was only 6,000 tons per annum.[30]

In Devon and Cornwall slates were quarried as early as 1296. In that year 23,000 'sclattes' were obtained from Birlond (ie Bere Ferrers) at 5d per thousand; and 10,000 were obtained from Hassel (probably either Hartshole or Parswell). These slates were used in buildings erected for miners at Martinestowe. Slates were also exported from this district, as 2,000 'sclat de Cornwayll' were bought for 10s for a house in the New Forest in 1363. The majority of the medieval roofing slates which have been examined are rectangular, with the sides more or less parallel, and the usual size is about seven inches by three and a half to four inches.[31]

The slates from the Delabole Quarries in Cornwall were famous from Elizabethan times; and Richard Carew, writing at the beginning of the seventeenth century, described them as 'in substance thinne, in colour faire, in waight light, in lasting strong, and generally carrieth so good regard as (beside the supply for home provision) great store is yearly conveyed by shipping both to other parts of this realm and also beyond the sea in Britaine and Netherland'.[32]

A comparison of Cornish slates with those of Pembrokeshire was made by George Owen of Henllys in 1602, when he said that they differed only in colour. The three colours of the Pembrokeshire slates were 'blacke, red and graye', and the black were so fine and smooth that 'they will serve to write on with any Botkinge' (bodkin). These slates could thus be used as writing-slates, or for sun-dials. Owen said that slate from North Pembrokeshire was sent by water to 'Harfors (Haverfordwest), Pembrok and Tenbye and to dyvers parts of Ireland and sould by the thowsand sometymes deere, and sometymes cheap as the plentie and scarcitie in those towns doe require'.[33]

Sion Tudur, who was Registrar of the Ecclesiastical Court at St Asaph about 1580, sent a poem to the Dean of Bangor asking for 3,000 slates to be brought down to Aberogwen where a ship would convey them to Rhyl. He prayed that the Dean might live 'three lives' and that there would be no broken slates in the consignment. The slates were needed to replace the thatch roof of Tudur's house, which was not weather-proof.[34]

Examples of the use of slate in the reign of Elizabeth include the slating of the chancel of St Asaph Cathedral, which was done under the orders of Bishop William Morgan (1541–1604). In 1682 other parts of this cathedral were slated with Penrhyn slates; and when in the 1930s the roof timber was being repaired, the slates were found to be strong enough to be replaced on new timbers. Samples of slates over 250 years old taken from the roof of St Asaph Cathedral can still be seen at Penrhyn Quarry.[35]

Various accounts of leases occur in the Calendar of Wynn of Gwydir Papers, 1515–1690; and in some there are references to slate. One, in the period 1568–74, records:

'I did demise unto Gruffith one tenement called Lloyn y bettws . . . condicione sequenti . . . 20s to me towards the slating of the dwelling house, and he to send the carriage of the slates and the meat of the slaters and carpenters.'

On 18 August 1597, among the memoranda of John Wynn, there is the order, 'slate the cattle-houses'. On 29 January 1656–7 there occurs this entry: 'Hearing, at Lady Anger's funeral that he has an inward bruise caused by a fall from his horse, Robert Mostyn sends to Richard Wynn at Plas Tirrion, a piece of Irish Slate, to be scraped and drunk in a posset before bed-time.'[36] This refers to 'alum slate' which was formerly used medicinally in the form of powder, and was in fact partly coal, partly alum stone and partly marcasite.[37]

During his travels in the years 1536–9, John Leland remarked that 'the houses withyn the town of Oswestre be of tymbre and slated'. Leland also noted that in 'Ise Dulesse' (Is Dulas, on the east side of the River Dulas in Denbighshire) 'they dig oute slate stones to kyver houses'.[38] In Wrexham slates were used to roof houses in Elizabethan times. One account gives the price for 6,000 as 40s, including transport.[39]

Glyn Ceiriog Quarry in Denbighshire was producing slates in the late seventeenth century: an entry in the Chirk churchwarden's account for the year 1675 reads: 'Pd. Nathaniell Roger for quarrying 3,000 slates from ye Glyn Quarry 15s od.'[40] There was also quarrying in Llansannan, west Denbighshire, in the early eighteenth century. An entry in the parish account books for 1737 states: 'Pd. for digging 2,000 slates 11s od.'[41]

Slate was used on the better type of house in North Wales in the sixteenth century, which was a period of considerable building activity. Examples in Caernarvonshire, noted in 1960 by the Royal Commission on Ancient Monuments in Wales and Monmouthshire, included a small two-storey house at Pant-Glas-Uchaf in Clynnog parish, which had a roof of rough early slates; a one-storey house with a roof of old slates at Merbwll in the parish of Llanaelhaearn; and at Glascoed in the parish of Llanddeiniolen a two-storey house of 1600 with a staircase projection which was slate-hung.[42]

Thomas Pennant in his tours in Wales around 1810 reported that 'a coarse slate' was found in the hills near Llanidloes in Montgomeryshire. In many regions, however, he said that 'shingles, heart of oak split and cut into form of slates', remained the 'ancient covering of the country'.[43]

In Scotland the earliest slate used seems to have come from the island of Easdale, off the coast of Argyll between Crinan and Oban. It has been claimed that the Falconer's Castle at Appin was roofed with Easdale slates in 1631; and Armaddy Castle, a seat of the Earl of Breadalbane, was covered with these slates in 1676. The Ballachullish Slate Quarries, situated on the shores of Loch Leven, consisted of two quarries about half a mile apart: the West Quarry was being worked about 1697, and the larger East Quarry was opened about 1780.[44]

Large pieces of slate, known as 'stone slates', were used for building and roofing in the eighteenth century in areas such as the upper part of the Conway Valley, the valleys of Machno,

Lledre and Llugwy, and the district around Llanberis and Beddgelert, where the rock was easily obtainable in pieces of large size.[45] The slate buildings usually had large stones at the base of the walls and smaller ones nearer the top.[46] The practice of cladding walls with slate to protect them from the westerly winds was noted in Caernarvonshire by J. Evans in 1812. Evans described how each succeeding row partially overlay the one below: 'Houses thus fronted with a fine coloured slate, the slates well selected, and put on neatly with black or dark grey mortar, assume a handsome appearance.'[47] Occasionally some of the walls thus protected had patterns of different coloured slates, and this added to their attraction. Patterned and plain walls are still to be found in parts of Caernarvonshire.

Certain rocks, such as shale, limestone and sandstone, can be split along the original bedding plane into sheets, the thickness being determined by the original beds. Such sheets were known as 'stone tiles' or as 'slates' and were used on roofs when genuine slates were not available. Stone tiles of Pennant Sandstone were used for the roofs of Roman villas at Ely and at Llantwit Major in Glamorgan; and in Northamptonshire 'Collyweston Slate' occurred in the Inferior Oolite Limestone Series. Likewise stone tiles, called 'Stonesfield Slate' from the village of Stonesfield in Oxfordshire, occur in beds of impure micaceous sandstone belonging to the Inferior Oolite Limestone Series. 'Cotswold Slates' occur in the Chipping Norton Limestone, where they are found between the Inferior and Great Oolites. All these tiles are thick and heavy, and have been used mainly for local needs.[48]

Dr E. A. Lewis has shown, by his analyses of the extant records of the ports of North Wales from 1562 to 1603, that at this date there was a small export of roofing slates from North Wales to Ireland. The coasting trade of north Caernarvonshire and Anglesey had fairly flourishing sea ports at Beaumaris, Caernarvon and Conway; by the middle of the sixteenth century, each of these had some mercantile shipping and long-established trading contacts with Chester and the creeks of the Dee. Most of the recorded trade was directed either to or from Beaumaris and Caernarvon; and the traffic with Ireland on the outgoing side, from 1583, consisted chiefly of the export of slates. In 1587 the number of slates exported amounted to 100,000, about one-tenth of the number exported to Ireland a century later. Over half the carrying on the incoming side was performed by Irish and Manx shipping but Lancashire and Cheshire ships often shared a substantial amount of the total traffic. The head port for North Wales was Chester; but there were customs officials at Beaumaris, Caernarvon and Conway.

In 1559 Wales had been brought into the English system of customs revenue; so that tunnage of 3s per tun was paid on all wine imported. In addition an *ad valorem* subsidy duty of 12d in the pound was paid on all merchandise imported or exported alike by aliens and denizens; and a further *ad valorem* custom of 3d in the pound was paid by alien merchants. Tunnage and poundage were levied in Beaumaris from 1577.

Most of the shipments were from Caernarvon or Beaumaris to Carlingford, Dublin, Waterford, Knockfergus, or ports vaguely designated 'Ireland'. The ships included the *Elizabeth* of Beaumaris, the *Jesus* of Ardglass, the *Edward* of Liverpool, the *Sondaye* of Strangford, the *Gallion* of Beaumaris and the *Bartholomew* of Chester. The tonnage of these vessels ranged from 5 to 24 tons.

The *ad valorem* prices give a rough indication of wholesale prices of slates during the second half of the sixteenth century: 'single' slates of small dimensions, about 10in × 5in, were 1s 8d per thousand, and double slates of larger size about 12in × 6in were 2s 8d per thousand. These slates were from ½in to ¾in thick. The numbers of slates leaving the ports were comparatively small: from 1 January 1584 to 28 August 1585, 48,000 slates; from 26 May 1586 to 18 July 1586, 86,000 slates; from 24 February 1587 to 16 September 1587, 100,000 slates; from 12 October 1589 to 23 July 1590, 66,000 slates.[49] At this date Beaumaris was the most important port of North Wales, and had control over the customs of the whole coast from the Conway to the Mawddach. By the late eighteenth century, however, it had been overtaken by Caernarvon, largely as a result of the increase in the slate trade.[50]

The North Wales Port Book for 1725–30, which has some gaps, shows that the trade of Beaumaris and Caernarvon was increasing, with Caernarvon in the lead. Comparatively small cargoes of slate were sent from Conway and Holyhead. Small vessels, however, and the fact that the harbours could not accommodate bigger ones, continued to limit the slate trade generally. Slates were sent chiefly to Dublin, Liverpool, and Chester. The vessels included the *Marygold* of Caernarvon (20 tons), the *Mayflower* of Beaumaris (14 tons), the *Thomas and John* of Bangor (12 tons), the *Blessing* of Holyhead (15 tons), the *Marlborough* of Caernarvon (15 tons), the *Sharke* of Caernarvon (7 tons), the *Speedwell* of Caernarvon (6 tons), and the *Boadicea* of Dublin (40 tons). Numbers of slates exported included 123,000 from Beaumaris between 31 August and 25 December 1729; 200,000 from Caernarvon between 1 and 31 October 1729; 29,000 from Beaumaris between 11 and 12 August 1730; and 197,000 from Caernarvon between 1 and 12 August 1730.

Slates exported from Conway included 25,500 from 18 July to 31 August 1729; and on 28 August 500 slates and 'a parcell of flagg stones' were sent from Conway in the *Darling*. Between 2 September and 21 October 1730 127,000 slates were sent from Caernarvon along the coast to Chester; and 195,000 were sent in the same period to Dublin. Shipments along with the slate included '10 dozen bottles Wines' in the *Dolphin* of Caernarvon bound for Chester on 5 March 1729; '660 Red Herrings in the *Mayflower* of Beaumaris bound for Chester on 27 November 1729; '3,000 Laths' in the *Hopewell* of Caernarvon bound for Holyhead on 6 August 1730; '2 Ca Welsh Ale' in the *Sinah and John* of Caernarvon, bound for Dublin on 10 September 1730; and '12 peggs Malt' in the *Morning Star* of Caernarvon, bound for Chester on 17 September 1730.

Between June 1729 and December 1730 over 2,500,000 slates were sent from the ports of Beaumaris, Caernarvon, Conway and Holyhead. Over a million of these were sent to Ireland, and the rest were exported coastwise. The number of vessels cleared out of Caernarvon in this period was 111 and the number from Beaumaris was 47. The average burden was 16 tons.[51]

2

PENRHYN QUARRY, 1731–1782

Until the second half of the eighteenth century all the Welsh quarries were small, shallow excavations worked by the local people for their own needs or for sale. Penrhyn Quarry was no exception, and the early accounts show that a small number of quarrymen were paying rent on the slates to the owners, who were making little investment in the enterprise. The earliest extant record seems to be of the year 1413, when an item in the rent-roll of Gwillym ap Griffith records that a number of tenants paid 10d each for working 5,000 slates. This was only half the rate charged in 1450 for the same parish of Llandegai, when 4d per thousand was paid to the first William Griffith.[1]

A reference to slate quarrying on the Penrhyn estate was made on 24 November 1544, when Sir Rees Griffith leased land in Maenol Bangor, the original manor of the bishops, to David ap Thomas, burgess of Beaumaris, with permission to raise slates 'in all places and quarell' in the parish of Llandegai, paying royalties (unstated) to Sir Rees, and with the right to ship the slates and other commodities from Aberogwen and Abercegin. David ap Thomas was given the monopoly of buying slates from 'all sellers' in Llandegai; and the annual rent for the lease was £12, payable to Sir Rees.[2] In David ap Thomas's will, dated 6 February 1568–9, by which time he was resident in Bangor, the only reference to slate occurs among money owed to him: 30s 4d for 13,000 slates by William Spen of Denbigh.[3]

Another early reference to quarrying at Penrhyn occurs in an account, undated but circa 1610, of the rents and profits due to Piers Griffith of Penrhyn for messuages and lands in Maenol

Bangor, Cororion and Llanllechid. This contains the entry: 'Cilgeraint and the quarries of slate there £6.'[4] Piers Griffith is traditionally reputed to have sailed with Sir Francis Drake at the time of the Armada, although he did not come of age till 25 September 1589; he is also said to have been prosecuted for illicit trading with the Spaniards. Whatever the truth of this, in James I's reign he was in serious financial difficulties and was obliged to mortgage the Cochwillan and Penrhyn estates. Evan Lloyd and Sir Richard Trevor bought the whole estate in 1616. The indenture, dated 4 December 1616, by which Piers Griffith disposed to Evan Lloyd, who had married Mary, the daughter of Sir Richard Trevor, the manor of Penrhyn for £9,000, mentioned the slates 'neare to the Churche of Llandegay . . . commonly called or known by the name of ye Felin (mill) Slattas'.[5] The estate was then sold in 1622 to the Lord Keeper John Williams, who was descended from the Griffith family of Cochwillan and Penrhyn; and he devised it to his brother's son, Griffith Williams. On the death of Griffith Williams's grandson Griffith in 1684, the estate went to his sister Frances; and it was divided at her death between her two sisters, Anne and Gwen. Anne married Thomas Warburton of Winnington in Cheshire, and Gwen married Walter Yonge of Escott in Devonshire.[6]

John Doulben was the agent for Penrhyn in 1718, when Sir Walter Yonge and Thomas Warburton were in charge of the estate. The 'whole profit of slates sold between 10 October 1718 till 8 November 1719' was given by Doulben as £182 14s 5d.[7] John Doulben's accounts from May 1731 to May 1735 are extant, and give the following decreasing quantities of slate sold: 1731–2, 1,140,400; 1732–3, 1,204,400; 1733–4, 811,160; 1734–5, 645,800. Some slates were sent by cart, but most of them were sent by sea to such places as Conway, Chester, Liverpool, Whitehaven and Dumfries. The vessels used included the *Betty* of Chester, the *John* of Conway, the *Hopewell* of Dumfries, the *Happy Lyon* and the Conway *Darling*. The 'landlord's proffit' for 1730–1 was 1s 4d per thousand slates.[8]

In 1736 John Doulben was succeeded as agent for the Penrhyn estate by John Paynter. By this date Thomas Warburton and

Sir Walter Yonge has died, and the estate was being looked after by Mrs Anne Warburton, the widow of Thomas, and the Right Hon William Yonge, the son of Sir Walter. Paynter's agency lasted eight years, from 1736 to 1744; and during this period he appeared to promote the interests of the quarryowners with considerable vigour. According to his accounts, the 'clear profit' on slate for the years 1737–43 was as follows:

	£	s	d
1737	93	17	1
1738	123	9	2½
1739	84	6	5
1740	80	16	8
1741	44	8	11
1742	74	18	7
1743	53	3	10

The slates sold between November 1743 and May 1744 were never accounted for.[9]

The payments to the carriers in these years were:

	£	s	d
1737	246	7	3
1738	324	1	9
1739	305	13	2
1740	293	0	5
1741	116	13	5

The number of slates produced in 1741 was 666,700, indicating a rent of 1s 4d per thousand to the landlords.[10] The numbers for the other years were not given; but on 25 October 1738 Paynter claimed in a letter that the tenants were having to pay 2s 8d per thousand to the landlords, so it is not possible to work out accurately the production of slates for Paynter's period of office. This letter, dated 25 October 1738, was sent to Sir William Yonge; and it explained that the 'Caernarvon slates', that is, slates from Cilgwyn Quarries, were 'a great hinderance to the sale of yours at

Aberkegin'. The name 'Aberkegin' or 'Abercegin' was given to a small inlet close to the town of Bangor, which in 1790 was converted by the first Lord Penrhyn into 'a commodious harbour'.[11]

Paynter asserted that the 'Getters and Carryors' of the slates from the 'Kilgwun' quarries, in the parishes of Llandwrog and Llanllyfni, paid no rent for their slates to anyone. This was accurate, as the Crown owned the land and did not bother at that date to collect any rents. In 1745, however, as Paynter predicted in his letter, John Wynn of Glynllifon, south of Caernarvon, took out a thirty-one-year lease of waste land, including that on which Cilgwyn was situated, at a rent of 10s per annum and one-tenth part of the clear profits.[12] John Wynn died in 1774; but his son Thomas, referred to in Paynter's letter, retained the lease until it expired in 1776. Thomas was MP for Caernarvonshire from 1761 to 1774, when he lost the election to Thomas Assheton Smith, accepting as consolation the title of Lord Newborough.[13] The Wynns did not work the Cilgwyn Quarries or collect rents from the quarrymen.

Paynter's letter is worth quoting at length. He began by stating that Cilgwyn slates were easier to obtain than those of Penrhyn; and he continued:

> When they are got they split them into any sizes they please, and make them much lighter, thinner and broader than your Quarrymen can possibly do. Notwithstanding all these Advantages in the *getting part* they sell every thous.d for Ten shillings; your poor tenants have but seven, your Farm being Two shillings and eight pence p. thous.d, and nine and eight pence is the greatest price that ever was had for a Thousand of Slates at Aberkegin. Add to all these a vessel that will Cary but 20 M (thousand) of yours, will load 25 M of theirs: then as to the selling part, I must acquaint you that a large ship is constantly kept to carry 'em to the London Market, an old Servant of Mr Thomas Wynn's is the Person to whom they are consigned at London, and one Mr Lewis Nanna, a topping tradesman, is the factor at Caernarvon. In like Manner they have trade with Dublin, for

the Small Slates; and there is not a town in England or Wales that have ever seen theirs, but prefer them to yours, nor a Vessell that can get a loading at Caernarvon that will come to Aberkegin, for they at Liverpool or Chester will very readily get 14 shill. per thous.d while we can sell for no more than 13s. And indeed the boats have dis-used this Place so long thro Mr Doulben's peevishness that 'tis with great difficulty I get any home trade at all, for which reason I was forced to solicit a trade with places where the Caernarvon slates are not known; and if you'll give yourself the trouble to run over Mr Clegg's letters you'll see the Prospect I have of succeeding and be able to judge what best to do and to advise. But should I contract with Mr Clegg for a certain number per annum, which is his and my drift to do, there's another block in my way and that is the Quarries are so monstrous hard, so full of Rubbish, and so much Water'd, that 'tis exceeding difficult to get quantities agreeable to the demand I expect as 'tis. Also to get Labourers to work; for all the old workmen had in a manner left off seeing that no body took care to promote a trade and betook themselves to Cobbling, weaving and other little businesses to get bread, so that one half of the Houses usually inhabited with Quarrymen are now held by People that are quite Strangers to the business. All those things put together induced me to think of Ingrossing the whole Slatory of Caernarvon into my own hands if 'twas possible, by contracting with all persons concerned to take all the slates they could get for a Certain number of years and so by that means to keep up the price of yours, by sending the others to Markets that shou'd not interfere.

This could it have been done would have added many hundreds to the Estate of Penrhyn. But all my Endeavours have proved inefectuall, and all I was able to do was to Engage two substantial Tradesmen of the Town to buy for me upon commission every slate that they suspected would be otherwise carried to such and such Markets for this one Summer. Mr John Wynn of Glynlleevon seemed inclinable to serve me in this treaty that I was in with the Quarry Men, and recommended it to his own Tenants, to agree with me (Tho I believe he did not consider what he was about). But 'twas all in vain, they would not be bound for any Consideration. So now Sir I am to tell you which way I think you may still effectually do the Business; if I think wrong, I hope you'll excuse me, for I have nothing more in view than increasing the income of this Estate. This *Kilgwun* Mountain is no Man's property whatsoever, nor does any Person claim the least priviledge from it, except the neighbouring freeholders who have a right of Commoning, I mean a right to the Herbage. Therefore, Sir, would it not be easy for you to obtain a Grant from the Crown of the mines and other Royalties in that Common?

I remember Sir G. Williams some years ago coming down from London, told me that you had procured him a grant in Denbighshire of all that was not already granted and that he looked upon it as no great favour; but that I afterwards found was not so, for Sir R. Cotton had the very thing and I believe got it without much difficulty, not above two or three years ago. If such a thing could be done in Denbighshire (a Mineral County) sure it may in Caernarvonshire; and if it can Sir I heartily recommend it to yr. Consideration, for I assure you 'twill be an Extraordinary advantage to this estate, such as I won't mention till I hear further from you. I know that some of our neighbours who expect to succeed in P-l-m- have an eye on this very thing. But as the getting it wd render them very unpopular (for they might thereby disoblige the freeholders) 'tis drop't, I verily believe on that account. Besides Sir, I should tell you that the very mountain in Llandegai Parish, which I allways took to be the Sole Property of Penrhyn (except a right of Commoning to that Scrub Tenem.t of Williams's) till of late; and now I find that the Bishop of Bangor claims the right there as do also the Inhabitants of Llanddeiniolen and that Mr Doulben in all his time never interrupted them. This surprised me a good deal, and put me upon Enquiry of all the ancient People.

Page 33 (*above*) Penrhyn Quarry in 1808; (*below*) Dorothea Quarry from the air

Page 34 Penrhyn Quarry about 1890, with Talcen Mawr in the foreground

They most of them can say that in Sir Robt. Williams's time no man durst as much as turn a sheep upon Llandegay Mountain nor take any other liberty whatsoever without his consent and that he annually, when in the country, walked the Boundaries himself.

Upon this I gave notice to all the neighbouring parishes that on Whitsun Monday last I would walk the boundaries of Llandegay; and accordingly I did, with some hundreds of People of all ages, threw down all inclosures and impounded all strange Cattle and did everything I could think of to assert the reputed ancient rights. Twas high time to stir, for the neighbours began to think and to say that Penrhyn had no power to build cottages upon the Waste, nor to get Slates any more than others, except in their own hand. But I hope I have convinced them to the contrary. However Sir, a grant from the Crown is an indisputable thing where there was none before, and would stop the mouths of all, whereas should there happen any Disputes or should any other person procure a grant of all not already granted it would certainly include Llandegay as well as other mountains, and you would probably be deprived the Priviledges both of getting slate there and of erecting cottages for the Quarrymen, for certain it is that no Gentleman or freeholder whatsoever has power without a grant, over any Adjacent Common except the Herbage, the Soil, Mines etc, being the right of the Lord only. Be pleased Sir, to favour me with your thoughts hereon and be assur'd I shall Always do the best I can.[14]

This letter reveals the long-standing dispute over the boundaries of the Penrhyn estate, which was to occur at intervals in the nineteenth century. It is also in the nature of a defence, for we learn later that during Paynter's agency the arrears of slate payments amounted to £861 7s 10½d. £302 7s 4d of this had been recovered, and a further £12 was still expected, making £314 7s 4d to subtract from the arrears, leaving £540 0s 6½d. Being subtracted from the eight years' profit, this left as the 'clear profit of the Slates for all the eight years of Mr Paynter's agency (when several schemes were used and great pains taken to encourage the Trade)' a sum of £8 0s 2d. Moreover, 'a Slatt reeve was paid in that time at £4 a year, £32, so that the Landlords lost in that time by the Slate trade—£23 19s 10d'.[15] In fact, slightly more than this may have been lost, as in 1737–8 the understeward and 'Slate Reeve' was paid £8 and the 'Slate Reeve at Aberkegin' received £5.[16] In 1748, however, the slate reeve was indeed paid only £4 as a year's salary.[17]

John Paynter, despite his good intentions, was by 1745 proved to have been something of a rogue. A letter of 7 November 1745, from his successor Robert Bridge to Mrs Anne Warburton in Chester, informed her that Paynter had left Penrhyn for Cheshire, leaving behind a debt of £880 to the owners, and that, since he was penniless, a bill in Chancery to recover the debt would only add to the loss. Bridge informed Mrs Warburton that nothing more could be done, and assured her that Mr Paynter had 'had no favour from me'. According to Bridge the only assets Paynter had left behind were 'a bundle of Iron hoops, an oak Table and some lumber not worth twenty shillings'.[18]

On 28 April 1746 Robert Bridge paid twenty-five quarrymen the money due to them 'after their last Account settled with Mr Paynter and before 1 May 1744'. The money was either paid to them at 3s 6d per thousand slates, or 'allowed in amount of their Rents' for their tenements, making a total payment of £22 3s 11d for 126,700 slates. Between 1 May 1744 and 1 May 1745 610,000 slates were sold, and the 'clear profit' at 1s 4d per thousand amounted to £40 13s 4d.[19]

During part of the period of Paynter's agency, between 1730 and 1740, larger sizes of slates were introduced in Penrhyn Quarry. Before that time the slates had been called 'Singles' and 'Doubles'; and 1,000 'Doubles' were considered equal to 2,000 'Singles'. Another kind was then introduced, which doubled the size of those called 'Doubles'; these were called 'Double doubles', and 1,000 of them equalled 4,000 'Singles'. General Hugh Warburton, Mrs Anne Warburton's son and the co-owner of Penrhyn, was said to have given a still larger size the name of 'Countesses', and to have called the Double doubles 'Ladies'. Slates of still greater size were called later 'Duchesses'; and those known as 'Queens' and 'Rags' were sold by weight. The slates were hooked on to the wattling of the roof by a peg driven through a hole near the head of the slate; and up to 1740 moss slates, generally not larger than 10in × 5in to 12in × 7in, were the ones usually produced. They were so called because it was the custom to push moss under the slates to keep out the rain and wind. Their thickness varied considerably, because the craft skill was not well developed; customers usually bought not by weight but by tally. The slates were sold per mille of 1,200 at the point of shipment; but sometimes 1,260 slates per mille was the rule, in order to cover breakages.[20]

Ladies were about 16in × 8in; Narrow Doubles were 12in × 4½in and Countesses were 24in × 10in. Slates over 30in long were called Queens. In the eighteenth century at Penrhyn Quarry, however, Countesses were only 19in × 10in.[21]

The aristocratic names of the slates often aroused interest and amusement, and a poem on the subject was composed by a North Wales county judge in the early nineteenth century. His comments on the new terms included the following lines:

It has truly been said, as we all must deplore,
That Grenville and Pitt have made peers by the score;
But now 'tis asserted, unless I have blundered,
There's a man that makes peeresses here by the hundred.
He regards neither Portland, nor Grenville, nor Pitt,
But creates them at once without patent or writ;
By the stroke of a hammer, without the King's aid,
A lady, or countess, or duchess is made.
Yet high is the station from which they are sent,
And all their great titles are got by descent;
And where'er they are seen, in a palace or shop,
Their rank they preserve and are still at the top.
Yet no merit they claim from their birth or connection,
But derive their chief worth from their native complexion.

And all the best judges prefer, it is said,
A countess in blue to a duchess in red.[22]

Other names for the different sizes of slate, however, were used towards the end of the seventeenth century; and Randle Holme in his *Academy of Armory*, published in Chester in 1688, listed some of these. They included 'Short Haghattes', 'Chitts', 'Jenny why jesteth thou' and 'Rogue why winkest thou'. It is likely that these slates came from North Wales; but the names may not have originated in the Welsh Quarries.[23]

After John Paynter's time the 'clear profits' of the slates from 1 May 1744 to 1 May 1747 were £69 13s 5d; but out of this had to be deducted £27 1s 4d for bad debts contracted within that time, and £12, the three years' salary to the reeve, leaving 'a bare profit' of £30 12s 1d. In 1747 Robert Bridge, the new agent, reported that the price of slates had been reduced in the previous year from 4s 10d per thousand to 4s 2d, of which reduction the quarrymen had been responsible for 2d per thousand. After this lowering of prices two sloops had come from Dumfries and 'holpt the Trade a little last year'; it was 'now worse than ever daily growing worse and will not probably clear £10 in 1747'.

Competition from Caernarvon continued to be blamed: 'The Caernarvon Slates (which takes away our trade) are lighter than the Penrhyn slates and consequently more is got by the freight of them and as they are softer and easier workt the Slaters recommend them tho' Penrhyn Slates are apparently better.' By this time John Wynn had secured his lease of the Cilgwyn Quarries, but 'the people that get the Caernarvon slates sell as they can sometimes higher and sometimes lower without paying any Merchants or profit to the Owners of the quarry, whereas the Penrhyn Slates are (now the price is lower'd) charged with a profit of ten pounds per thousand and the price is fixed and unalterable'.

Robert Bridge suggested that if the landlords would allow the tenants to sell their slates on their own account without payment or rent, it would restore the trade and enable the tenants to pay the rents of their cottages on the common. Some of the tenants rented land from £3 to £10, and also sold slates, and with more sales, Bridge claimed, they would bring more money into the estate. The agent proposed that regulations should be imposed on the tenants selling on their own, as an experiment for three years, the landlord in this case being given £6 per thousand of all slates sold. If this did not answer, then the tenants should, he thought, be allowed complete freedom to sell the slates without any payment to the landlords. He thought this would increase trade, or at least reduce the number of bad debts.[24]

During the year 1 November 1742 to 1 November 1743, £53 3s 10d was paid to the landlord; and £139 12s 5d was paid to the carriers. The total output for this year was over 797,000 slates; and the vessels used were the *Sherk*, the *Speedwell*, the *Dispatch*, the *Susana*, the *Mary*, the *Lyon*, the *Lamb* and the *Mayflower*. Small quantities were sent by the 'Garth boat', and by 'Mrs Evans of the Ferry'; and some were sent by cart.[25]

Despite the fact that Cilgwyn could undersell Penrhyn slates, the quarries were not worked systematically. The early methods of quarrying and transport were later recalled in these terms:

> The stones were smashed up with a hammer without any rule; any piece of rock which happened to be anything like a slate was made use of, but at least half of the good rock was thus destroyed . . . The men at that time took contracts from the landowners, paying two guineas royalty per annum for the privilege of quarrying the slates. Each man dug a hole where he thought proper, and many of the quarries to this day are suffering from the heedless way in which the holes were excavated and the rubble disposed of, usually thrown out close to the hole upon the good vein . . . The slates were transported from the pits upon iron hooks, laid across horses' backs, to a place called Voryd, upon the sea-shore below Nantlle, where the old rubbish-heaps are still to be seen . . . The rubbish from the pits was conveyed away in wheelbarrows and in bags upon the backs of men.[26]

In the 1750s the number of quarrymen on the Penrhyn estate

continued to be very small, although the output of slate was comparatively large. From 19 February 1752 to 23 April 1753, the quarrymen produced 2,290,300 slates, for which they were paid a total of £400 16s od, at 3s 6d per thousand. As the landlords then sold the slates at 4s per thousand, their profit for that period was £57 5s 9d.[27] From 19 April 1758 to 19 November 1761 there were only 36 quarrymen at work; they 'shipped and delivered on oath' 1,794,200 slates, for which, at 3s 6d per thousand, a total price of £313 18s 8d was paid. These slates were sold at 4s per thousand, giving a total of £358 18s 6d, and a profit to the landlords of £44 19s 10d.

By this date Robert Bridge had been succeeded by Richard Hughes. Hughes inherited arrears of £37 12s 9d, which had been due since Bridge's time and were now 'entirely lost'. On a 1758 consignment of 40,000 slates at 8s, delivered on board *Penrhyn* to Hugh Jones of Chester, there was a memorandum that the 'Accountant Hughes' had paid the quarrymen 7s per thousand for these slates, and had allowed the landlords 1s per thousand profit, which was included in the total.[28] In 1771 there were 25 tenants paying for tenements and quarry rent, and disbursements and payments were made by the 'late Richard Hughes Esq and continued by his son Hugh Hughes Esq'.[29]

Penrhyn Quarry was not opened until Richard Pennant (?1737–1808) took over the whole estate in 1785. Before then, slate was raised from two places on the estate: from the common at Chwarel Goch, east of the present quarry, enclosed by 1800, where red slate was got by the Penrhyn tenants for repairing and slating their farm buildings; and from above the present quarry, on the common called Frith. The present quarry was originally in a field known as Cae Braich y Cafn, and the early quarry was called after this field.[30]

Little investment in slate was made by Sir William Yonge of Devon or by Colonel (later General) Hugh Warburton of Winnington in Cheshire; but both received reports from the agents, and Colonel Warburton had an official interest in North Wales as Chancellor and Chamberlain in the counties of Anglesey, Caernarvon and Merioneth. Each of them received a large order of slates in London in 1750. Colonel Warburton's order, sent on 18 June 1750, consisted of 13,500 slates at 8s per thousand, and '2,000 Ladies or Double doubles at 16s'; this shows that by 1750 the new titles for slates were just becoming generally accepted. Sir William Yonge's order, on 3 July 1750, was for '44,000 Doubles at 8s, 3,000 singles at 4s, 4,500 Ladies at 16s, totalling £21 16s od'.[31] These orders suggest that the joint owners were taking a hand in the marketing of the slates; but as absentee landlords, they undoubtedly played a part in allowing the boundaries of the estate to become confused, with the result that many disputes had to be fought later to establish or re-establish these.

Sir George Yonge, son of Sir William, was joint owner of Penrhyn estate in 1757; and he and General Warburton were giving orders for slate in that year.[32] On 13 November 1765, however, General Hugh Warburton's only child and heiress, Anna Susannah, took an action of resounding consequences, not only for the Penrhyn estate, but also for the development of the North Wales slate industry: she married Richard Pennant, the son of John Pennant, a merchant in Liverpool. John Pennant was descended from the Pennants of Downing and Bychton in Flintshire, and was related to Thomas Pennant, the traveller and naturalist. He was immensely rich, having inherited sugar plantations in Jamaica from his father, grandfather and brother; and on his marriage to Bonella Hodges in 1734, two estates in the parish of Clarendon in Jamaica were merged. He continued negotiations for the purchase of the moiety of the Penrhyn estate held by the Yonge family, which had been begun as early as 1737; but these negotiations were not completed until four years after his death in 1781. In 1785, since Richard Pennant's wife already possessed the other moiety held by the Warburton family by right of inheritance, the estates of Cochwillan and Penrhyn became reunited.[33]

As early as 1 September 1768, however, John Pennant had taken over from the Yonges to the extent of issuing jointly with General Hugh Warburton 54 leases (48 in the parish of Llandegai and 6 in the parish of Llanllechid) to various tenants at differing annual rents, for twenty-one years, with a covenant that the lessee, for a further sum of 20s a year, should have liberty to work, or to find one labourer to work, in or about the slate quarry in Llandegai. These leases are all printed, and the tenants' rents varied from £13 to £1 6s. The majority of the tenants were designated 'yeoman', but there were also labourers, widows and at least one blacksmith.[34] This money payment was instead of the customary rent of an eighth of the produce which had been paid to the landlords, and was an improvement from the lessees' point of view. The slate was quarried for sale on the then common above the present quarry, which was enclosed in 1800; but output was small, since the groups lacked capital and could not afford to carry out unproductive work such as clearing.

An undated set of rules 'to be observed in the working and management of the Slate Quarry belonging to the Honble. Hugh Warburton and John Pennant Esq' was issued about 1770. The workmen were seventy-seven in number, so they included others besides the fifty-four local tenants who had been granted leases in 1768. The basic principle of organising the quarrying among groups of workmen, which became a feature of the industry, was here recognised in the rules:

> That the workmen seventy-seven imployed in the said Quarry do form themselves into Sets or Companies to consist of five, six, seven or more Men according as the places where the Slates are to be raised will admit of a convenient number to work . . . That each Set of Workmen do raise and dress the Slates in the most proper and beneficial manner according to such usual Sizes and Dimensions called Singles, Doubles, Ladies and Countesses as the several parts of the Quarry will admit of.

The details of the financial arrangement reveal that the men were to pay for the carriage of the slates, or to transport the slates themselves to Abercegin, where they were to be placed ready to be shipped. The slate reeve was to keep an account of these slates and was to 'sell or dispose of such Slates for the best

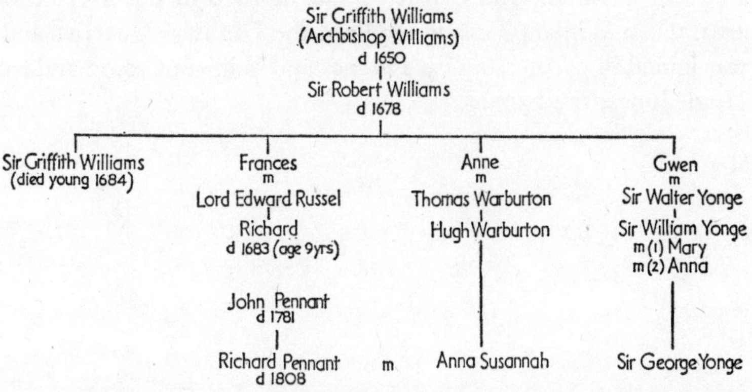

prices', and at the time of loading was to 'take a particular account of the Slates belonging to each Company'. The slate reeve was to present his accounts three times a year, in June, September and December, to Hugh Warburton and John Pennant, who would then pay him. On receipt of this money the reeve was to pay 'and divide the same amongst the Workmen according to the Quantity and Value of Slates sold and delivered by each Set or Company'. The final rule, referring to disputes, seems to lay down the principle of the owners' absolute authority, which was closely adhered to by Lord Penrhyn in the dispute of 1900: 'That if any Difference or Disputes shall arise between the Workmen or between them and the Slate Reeve relative to the premises the same shall be determined by the Agents of the said Hugh Warburton and John Pennant for the time being with a power for any of the parties to appeal to the said Hugh Warburton and John Pennant, whose determination shall be final.'[35]

The unorganised exploitation of the slate rock was effectively ended in 1782 in the Bethesda area, when Richard Pennant, the first Lord Penrhyn, took over the working of the quarries; in 1787 in the Llanberis area, when a private partnership obtained a twenty-one-year lease of the quarries in the Manor of Dinorwic; in 1800 in Nantlle, when John Price, Thomas Jones, Richard Roberts and John Evans were granted a lease by the Crown of

Cilgwyn Common and quarries; and in 1800 in the Ffestiniog area, when William Turner acquired the Diphwys Quarries and was joined in partnership by Thomas and William Casson and by Hugh Jones, the banker.

3
PENRHYN AND DINORWIC, 1782–1831

After the death of John Pennant, the Penrhyn estate was vigorously developed by the dynamic personality of his son Richard, who was created Baron Penrhyn of Penrhyn, county Louth, in September 1783.[1] He controlled the whole estate after 1785, with the assistance of his agent, William Williams of Llandegai, who was devoted to the wellbeing of the family. Williams died in July 1817, and his obituary notice recorded that he was in his eightieth year, and was the author of *Observations on the Snowdon Mountains*. The account continued:

> Mr Williams had been for many years one of the agents of the Penrhyn estate. He, in the year 1782, was the humble means of inducing the late Lord Penrhyn to 'form the wise and benevolent resolutions of opening a spacious slate quarry', at Cae braich y cafn in the parish of Llandegai. When his health began to decline, the reward of his long and faithful services was a handsome annuity.[2]

Some of William Williams's accounts give us a fair indication of the intense activity that went on during the years of control by the first Lord Penrhyn. There are lists of payments made 'on Acc.t of the Slate Quarry from October 1782 to November 1783', which include:

22 February Paid Harris to pay the Men for opening the bottom of the Quarry to get to the joints of the rocks.	£2 18s 10½d

14 March Paid Harris to pay several labourers for clearing a Body of Rubbish in the Quarry which was set to the Lab.s as a job.	£30 0s 0d
5 April Paid Labourers and Quarrymen for work done in the quarry.	£8 13s 3d

A significant entry is the one on 1 July 1783 which says, 'Paid partners for their Quarry adjoining the new Quarry—£72 0s 0d.' This refers to the quarry on the Common above the present quarry where the tenants had paid 20s a year for the privilege of quarrying slate on their own account, and shows that after 1783 this right no longer existed.

There are two entries in 1783 referring to provisions for the workmen. That for 1 November reads: 'Paid Heald at different times subsist (subsistence) and in corn to the Labourers and Quarrymen—£18 19s 2d.' Two days later we find: 'Corn had by some of them (labourers) £9 8s 3d.' The sum paid for making slates, that is the wages bill, for 1783 was £254 11s 2d; but there were also expenses to be met for making the road to the quarry, which in 1783 were £88 3s 1d. The amount received for the sale of slates in this year was given as £597 9s 2½d; and the carriage of slates was paid for at 6s per ton. From 21 August the carriage rates for Countesses and Ladies were lowered, as they appeared to be lighter than had been thought. Countesses were reduced from 12s to 9s 6d per thousand, and Ladies from 8s to 6s per thousand; but the Doubles remained the same, at 4s per thousand. The prices of slates at Abercegin in 1783 were:

Countesses	£1 12s 0d per thousand	
Ladies	16s 0d ,, ,,	
Doubles	8s 6d ,, ,,	[3]

William Williams, reviewing the work in progress in a letter dated 14 August, said that there were about 30 men at work on the road to Abercegin, and that there was 'a pretty good No of slates in the Quarry' which were 'much wanted at Abercegin'. Williams was intending to send off a cargo of 15,000 Countesses to Liverpool. He was also giving his mind to the question of enclosures on Llanllechid Common; and he saw the difficulties arising from the different qualities of soil and of situations. He thought the common worth enclosing, and suggested that 'strangers' should be allowed to enclose on the Penrhyn boundaries; but he realised that there would be objections if their allotments were fixed higher up the mountain. On Llandegai Common, where some enclosures had to be re-made because of encroachment, these enclosures were proceeding more smoothly and only needed approval from Richard Pennant.[4]

Richard Pennant was one of the first landowners to turn his mind from agricultural improvement to the exploitation of the slate resources on his estate. He promoted agricultural improvements; but his outstanding achievement was his investment in slate, which served as an example to others. He was elected MP for Liverpool in 1767, and he was re-elected in 1768, 1774 and 1784.[5] Not surprisingly, Liverpool and Runcorn were the chief ports in his day for Penrhyn slates. Part of the family's income was derived from estates in Cheshire, and part from their sugar plantations in Jamaica.[6]

It was in 1784 that Lord Penrhyn, as Richard Pennant had then become, secured a lease from the Crown covering the hundred of Uchaf, which comprised the parishes of Llandegai, Llanllechid, Llanfairfechan and Dwygyfylchi. He also secured a lease from Bishop Warren of the Penybryn foreshore for building the quay at Port Penrhyn; and he exchanged land with the trustees of Beaumaris Grammar School, so as to receive Dyffryn Mymbyr. Lord Penrhyn was also taking an active interest in his Jamaican estates, inherited on the death of his father in 1781. On 26 November 1782 he wrote to his agent Mr Falconer:

> Pray to let me know whether Slates would not be usefull in Jamaica, instead of Shingles, or Tiles; and as I have a large slate quarry, it would answer very well to me, to supply any quantity. In Kingston, and in Towns, I should think they would answer particularly well.[7]

There is no record of sales to the West Indies at this date, but in 1810 Thomas Pennant said slates were shipped for 'even the West Indies'.[8] The Pennants certainly cherished their slaves, albeit regarding them as a race apart, as is revealed by Richard Pennant's statements on 25 March 1783: 'I am glad to hear the Negroes are well. The Hearing a good account of them, and of the Cattle, always give pleasure.'[9]

The petition for the lease in the parish of Llanllechid was made on 19 November 1783, and the Lords Commissioners of the Treasury were addressed thus:

> That your memorialist having a considerable estate in the said parish of Llanllechid and in other parts of the said County of Caernarvon within the Hundred aforesaid, and having an intention of occasionally residing there, it would be an accommodation to your memorialist to have a lease of the said Hundred and Manor with the Royalties thereto respectively belonging, and a power of mining in the Commons, or Wastes . . . it being your memorialist's design to dig and search for mines and to work the same, in Case upon Trial it shall appear adviseable to do so.[10]

The petition was granted on 2 August 1784 'by and with Advice of our Right Trusty and Wellbeloved William Pitt first Commissioner of our Treasury'.[11] There were two slate quarries in Llanllechid, almost opposite each other, called Bryn Hafod y Wern and Moel Faban, and these were worked by the Pennants; but it was later proved that the lease did not give them the right to dig for slates, and a new lease had to be acquired.

Port Penrhyn, formerly called Abercegin, close to the town of Bangor, was naturally only a small inlet; but it was converted in 1790, by Lord Penrhyn, into 'a commodious harbour, capable of admitting vessels of 300 tons burden, for more conveniently exporting the slates from his quarries, about six miles distant'.[12] In his tour through North Wales in 1798, the Rev J. Evans visited Port Penrhyn. He noted that there were vessels there for London, Bristol and Liverpool, but that the chief trade was with

Ireland. He described the slates as being of 'all sizes, from large tombstones and slabs for pavement, down to the smallest size used for roofing'. He gave the sizes and prices, saying that the slates were sold 'for money only'. Later an elaborate system of credit and discount was developed. In 1798 the prices were higher by about 25 per cent than in 1783:

Inches		£	s	d
24 × 12	Duchesses per 1,000 (1,260)	3	10	0
20 × 10	Countesses ,, ,, ,,	2	0	0
16 × 8	Ladies ,, ,, ,,	1	0	0
12 × 6	Doubles ,, ,, ,,		11	0
10 × 5	Singles ,, ,, ,,		5	0
Various sizes	Patents per ton	1	6	0
,, ,,	Rags ,, ,,		18	0
,, ,,	Kiln Ribs per yard		0	3

By this date the factory for manufacturing writing slates had been established near the port; and Evans claimed that Lord Penrhyn was 'able to undersell the Dutch', while producing better quality.[13] Pennant, however, claimed that it was the Swiss who had previously supplied these slates. About 1809 he said that 136,000 writing slates had been exported.[14]

What had been, according to Thomas Pennant, 'the most dreadfull horsepath in Wales', was made into a good road by Lord Penrhyn during 1790–1 to open up Nant Ffrancon. This road, known as 'the old road', can still be seen going along the western side of the valley past Maes Caradoc and Blaen y Nant. In 1800 Lord Penrhyn continued the road to Capel Curig, where he built an inn. In 1802 the Capel Curig Turnpike Trust built a turnpike road, which went along the eastern side of the valley on a route practically the same as that followed by the present road. The earlier road represented the first attempt to open a way into the county from the east.[15]

The railway from Penrhyn Quarry to Port Penrhyn, a distance of 6 miles, was completed in 1801. The gauge was 1ft 11½in, and the horse-drawn trains consisted of 'small waggons, linked together in succession'. Two horses drew twenty-four waggons one of the five stages six times a day, carrying 24 tons each journey. Williams noted that 'the expedition and facility' with which the slates were brought to the ships was 'very remarkable'.[16] E. Pugh, in his earlier account, mentioned the railway's three inclined planes, asserting that the 'declivity' was almost 300yd. According to Pugh, the expense of making the railway was £5,000; but the savings effected were considerable. When the slates were brought down by cart, the cost of carriage was 5s per ton (a conservative estimate); but the railway had reduced the charge to 1s per ton.[17]

Lord Penrhyn died in January 1808, and obituary notices listed his achievements:

> In the course of a few years, the agriculture of his own large possessions, and of the neighbouring county, made greater advances towards improvement, than it had done for several preceding centuries. It was thus, that by the creation of an active and extensive traffic, the materials of which were drawn from his own estates, employment and food have been given to thousands. It was thus, that by his countenance and assistance, the example was set of making this county accessible to itself and to the world, by opening of roads throughout almost every pass among its mountains.[18]

A monument to Lord and Lady Penrhyn was erected in Llandegai Church with money left by Lady Penrhyn, who died in 1816. This monument is of marble, and represents two figures: a quarryman leaning on his crowbar with a slate knife in his hand, and a girl weeping for the loss of Lord and Lady Penrhyn. There are four medallions, which record the progress initiated by Lord Penrhyn. The first shows a child on a mountain-side feeding a goat. The second shows a child using slate tools in the quarry. The third shows a child teaching another to read the Bible. The fourth shows three boys with sickles in a field of wheat.[19]

During the depression of the 1790s, Lord Penrhyn set some unemployed quarrymen to work on his estate. As a result most of his land was enclosed; and waste land was drained, laid down in

Page 51 (*left*) Dinorwic Quarry, showing an aerial winch and two sheltering sheds; (*right*) Dinorwic Quarry, showing some of the galleries with Llyn Peris below

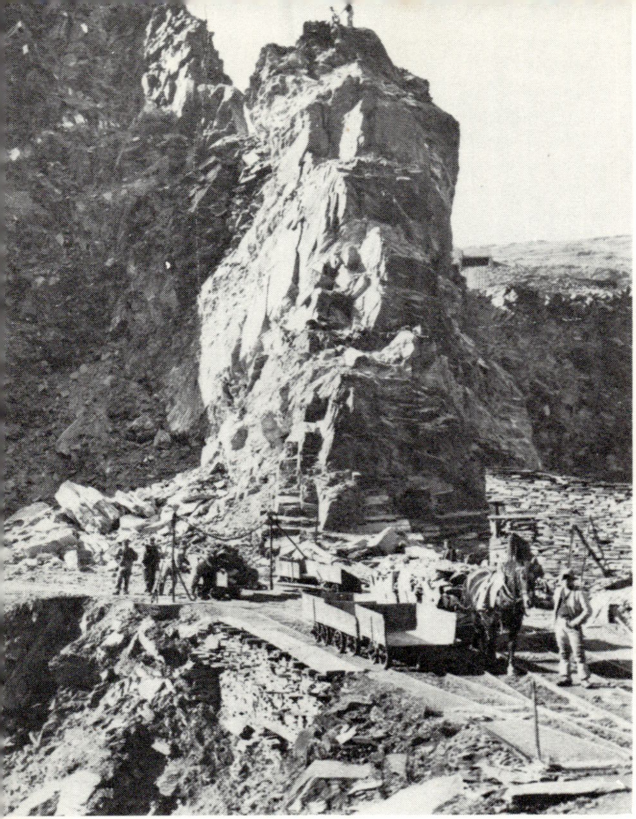

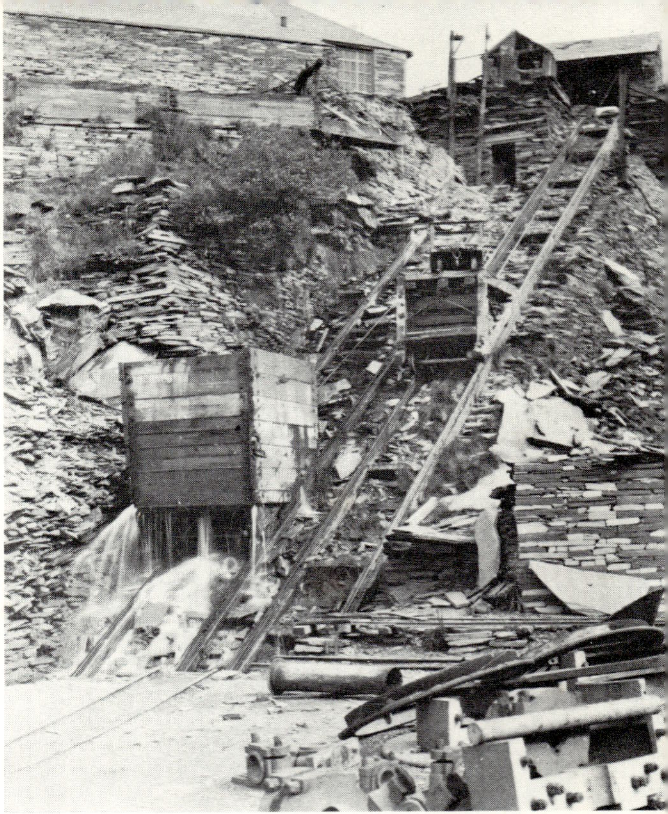

Page 52 (*left*) Votty and Bowydd Quarry, showing horse-drawn slate wagons; (*right*) the Votty and Bowydd Quarry water balance in use in 1936

grass, and let as leys to feed the cottagers' cows. Between 1790 and 1800, forty cottages were built: these contained sixty-three dwellings, 'many of them having double apartments'. These improvements, it was claimed, gave new life to 'a most dreary and forlorn waste'.[20] In 1798 the Rev W. Bingley observed on Lord Penrhyn's estates a method of fencing which was new to him. 'The fences were made with pieces of blue slate . . . driven into the ground about a foot distant from each other, and interwoven near the top with briars, or any kind of flexible branches to hold them together.' Bingley remarked that 'in point of ornament' they were 'sufficiently neat'.[21]

The other monuments left by Lord Penrhyn included a cottage named Ogwen Bank, which was about half a mile from the quarry. This was described in 1821 as 'an elegant and romantic little retreat—a perfect *bijou*—designed by the late Lord Penrhyn, and still used by the family, as an occasional resort for refreshment and recreation, on paying a morning visit to the Quarry, or other objects of curiosity'. Nearby was Pen-issa-nant, which had been intended as a dairy and poultry farm for Penrhyn Castle; and these two houses were open 'to all respectable visitors'. Inside Penrhyn Park there was 'a large private sea-water bath, with warm baths contiguous to it'. This was said to have cost £4,000; and it shows that Lord Penrhyn subscribed to the increasingly fashionable belief in sea-bathing.

In 1821 the castle itself was the one erected by Lord Penrhyn from the designs of the architect Samuel Wyatt, the brother of the quarry agent. The agent's residence at Lime Grove in the lower part of Penrhyn Park had also been planned by Samuel Wyatt; and it was considered 'more creditable to his taste than the greater work at Penrhyn Castle'.[22] In 1827 Thomas Hopper designed the present Penrhyn Castle for George Hay Dawkins Pennant, Lord Penrhyn's cousin. His plan preserved the remains of the original building and retained the fourteenth-century hall. Wyatt's unpretentious house was replaced by a pseudo-Norman castle with a keep rising to the height of 115ft. Douglas Hague declared that the building was 'theatrical' and 'about as homely as a great railway terminus'.[23] Charles Dickens in *All the Year*

Round referred to it as 'the sham baronial castle' inhabited by the 'Arch-magnate, the well-known Colonel Slater'.

Three years after Lord Penrhyn's death, the son of his faithful agent William Williams began a correspondence with the Lords of the Treasury under the assumed name of 'William Trotter'. In October 1811 he declared that 'the Slate Quarry worked by Lord Penrhyn many years in the parish of Llandegai is the property of the King unless a Grant has been made to Lord Penrhyn for it'. In December 1811 James Greenfield, the manager of the quarry, wrote to Lady Penrhyn:

> Your Ladyship will observe how basely William Williams Junior have acted in this Business, but it would perhaps be advisable not to let your knowledge of this circumstance transpire in Wales as we have got the information privately through Friendship from the Office.[24]

On 18 January 1812 Matthew Wyatt declared Williams Junior to be 'drunken beyond all decency'; and it was said that he had thrown away £1,000 in 'drink and negligence'.[25] His statements, which were calculated to damage the Penrhyn family, led to a thorough examination of the grant of Uchaf to Lord Penrhyn; and evidence about the ownership of Llandegai was submitted to learned counsel. On 21 November 1817 the Penrhyn family, now controlled by George Hay Dawkins Pennant, obtained a second grant. This was valid for thirty-one years; and unlike the first it gave them the right to search for and get slates within the parishes of Llanllechid and Dwygyfylchi. The memorial of 1811, which had claimed that the Penrhyn Quarries in the parish of Llandegai belonged to the Crown, was dismissed. No further proceedings were taken; but the case had raised some doubts and difficulties.

In 1821 Dawkins Pennant sought to prevent any competitors from making slates on the wastes of Llandegai and Dwygyfylchi; by acquiring a lease of Hirael Quay at Bangor, he obtained control of all the shipments of slates from the district of Bethesda.[26]

After assertions that the Pennant family had no right to work quarries in Llanllechid, a Parliamentary Enquiry was set up. Reports were submitted by John Atkinson, James Walker and others. A report from W. Hazledine to Dawkins Pennant said that there were twenty-five men at work at Bryn Hafod y Wern, the most important quarry, and that the work was being carried out by James Greenfield in a 'masterly' manner. Hazledine, however, urged Dawkins Pennant to abandon that part of the Crown land, as he saw little prospect of finding the valuable Penrhyn vein there. A report from James Farey was condemned by the agent Benjamin Wyatt, as it was unfavourable to the Penrhyn family.[27]

On 22 November 1823 Joshua Trimmer, a tile and brick manufacturer of Middlesex, applied to the Crown for a lease to work the quarries in Llanllechid. Counsel declared that there was evidence of 'a certain kind of enjoyment on the part of the Crown by its Officers in ancient times'. They continued, however:

> Considering the unmolested enjoyment for the last thirty years by Lord Penrhyn of property which has been gradually advancing into such considerable value by the labour and expense and exertion of his lordship without any claim on any part of the Crown, any attempt now to disturb the rights of his Devises and to deprive them altogether of these improvements and of property, in a manner created by him, will be very ill received, and every presumption be raised against it.[28]

On 31 August 1824 Thomas Stanton sent his report to the Commissioners of Woods, Forests and Land Revenues: he had failed to discover the 'valuable Vein of Slate . . . so nearly adjoining', and was compelled to report that 'Mr Pennant had fulfilled the condition of his lease'.[29] In October 1825 Trimmer was granted a lease to work the quarries in Llanllechid.

The profits of the large Caernarvonshire concerns, Penrhyn and Dinorwic Quarries, were rising fairly steadily. The accounts in the Penrhyn Quarry Books for the years 1816–18 and 1820–9 give a useful summary of the profits being made:

Year	£	*s*	*d*
1816	9,802	15	3
1817	12,570	5	2
1818	13,990	13	11
1820	14,910	11	3¼
1821	16,398	3	10½
1822	25,850	19	9½
1823	30,026	0	5¾
1824	32,057	9	4
1825	23,160	2	5½
1826	29,992	11	6
1827	34,303	10	10
1828	22,681	16	2
1829	24,945	1	5 [30]

The early capitalist stage of Dinorwic Quarry began when the old quarries opened by groups of quarrymen on the mountain-side in the parishes of Llanddeiniolen and Llanberis in Caernarvonshire, belonging to Thomas Assheton Smith, the lord of the manor, were taken over, in 1787, by a partnership of Thomas Wright, Hugh Ellis and William Bridge. This partnership, which included two lawyers, worked the quarries for twenty-one years at a rent of £12 per annum. Ellis died in 1807, and in 1809 Thomas Assheton Smith took over control himself. A map of 1771 shows only two quarries worked as such, these being Allt Ddu and Bryn Glas; but there were other small quarries which had been worked very early in the century by quarrymen who paid a nominal rent to the lord of the manor. In 1788 a 'Great New Quarry' was begun by the partners, above Llyn Peris; and this, incorporating some of the old quarries, became the huge quarry on the north-eastern side of the lake, which extended over 1,400ft up the mountain.[31]

Wright, Ellis and Bridge made successful trials for slate in 1787; and between 30 October 1787 and 22 August 1788, they sold a large quantity:

171,000 Countesses	at £2 2s per thousand	£309	2s
151,750 Ladies	at £1 1s ,, ,,	£159	6s
125,000 Doubles	at £1 11s ,, ,,	£68	15s
456 tons 10 cwt Ton slates (sometimes called Rags) at £1 2s		£502	2s
		£1,039	5s [32]

Between 15 August 1787 and 4 February 1788, their expenditure was £769 17s 6½d. This included £84 4s 4d for clearing the 'best Quarries', which had been opened, but had been filled again; £352 4s 11d for paying sixty workers for raising and dressing slates in the quarries which had been cleared; £114 10s 6d for carriage, by sea and land, of the slates; £10 18s 0d to a quarryman named J. Griffith 'for setting Bargains to Quarrymen and for looking after their Work, 27 weeks at the rate of £21 per annum'; £1 5s 0d to John Pritchard, a quarryman at the large 'Old Quarry' on the common, for clearing some of the rubbish in this quarry, the payment being intended as an inducement to him to give up 'that Quarry which he had been in possession of as in his own Right for about two years before'; 4s 0d to 'two Bailiffs from Caernarvon for putting a Stop to a parcel of the Quarrymen opening a quarry in the Common which they had begun to do as in their own Right'; £3 6s 11d for repair of the roads, 'Ale given to Quarrymen', and for messengers and postage; and £6 to 'Thomas Assheton Smith Esq. for half a year's Rent of the Slate Quarries to February 1788'. In February the concern remained 'in Debt to the Adventurers' to the amount of £565 4s 11d, some £204 12s 7d having been received mainly from the sale of slates. The partners employed the local quarrymen wherever possible, but out of the charges £70 'at least' had been paid 'on purpose to prevent the Workmen going to ye Common to open and work Quarries for themselves'. There was also a large stock of slates at Caernarvon and at Moel y Don, about three miles south-west of Bangor and now called Port Dinorwic; and there were also slates remaining at Waen Wen, Cwm y Glo and some of the quarries, 'for want of being carried;

the Roads being excessive bad and the Carriers' Horses very poor'.[33]

Comments on this account, made by the three partners on 4 February 1788, underlined the need for cheaper transport if the concern was to succeed:

> It is to be observed that slate being so heavy an article makes profit to the Adventurers very trifling so as not to be an object even when there is a great Demand for them as there is at present, unless great quantity can be raised and brought to the shipping place which cannot be done without first laying out a very considerable sum in clearing the Face of the rock and making New Roads and repairing others so as to reduce the Rate of Cost as much as possible; which for some time past before the new Undertaking had been estimated at or near half the value of the Countesses and always half the Value of the Ton Slate, Ladies and Doubles, tho' at other Quarries the Tons etc are carried for less than the third of their Value and the Countesses for about a fourth of the Value. The distance from the Upper Quarry to the Shipping Place is upwards of eight miles, and no road that a Cart can go upon within three miles of the Quarries and the Roads where Carts do go upon in Llanddeiniolen and Llanfairisgaer to Moel y Don are narrow, rugged and dangerous and also very steep in many Places, and the Road to Carnarvon the same. To remedy these inconveniences as much as possible a New Road from the Quarries to Llanberis Lower Pool is begun at a considerable Expense, the Length of which will be one Mile and a quarter, and it is also proposed to assist the Parishes in repairing the Roads along which the Slate must be carried, and in making new ones where they can be shortened or Draught eased, especially near Moel y Don and also at Cwm y Glo, if leave can be had to make a Road thro' Ld. Newborough's lands.[34]

In 1788 the rates paid to the Dinorwic slate carriers, who were to 'put Slate on Ship Board or pay Mr Smith's Shipper', were as follows: 6d per thousand, Countesses; 6d per ton, ton slates; 4½d per thousand, Ladies; and 4s for every 20,000 Doubles. From Bryn Glas and Allt Ddu Quarries via Waen Wen to the sea, the rates were: 14s per thousand Countesses; ton slates, Ladies and Doubles 'after the Rate of 10s per Ton Quarry Weight'. From Cwm y Glo and Pen y Llyn the rate was 5s 9d per ton 'for all kinds of slate'.[35]

In November 1788 the partners had gained permission from Lord Boston, who owned land in the parish of Llanfairisgaer, for a new road from Llanddeiniolen to the ferry at Moel y Don. The old road had been 'indicted for being out of repair'; and since the new road would principally benefit Thomas Assheton Smith and 'the occupiers of his Slate Quarry' they would 'subscribe liberally towards completing the Work'. The road, when made, was expected to 'greatly ease the Farmer in bringing up his manure as well as the Slate Carrier and Traveller'.[36]

The end of the eighteenth century was the beginning of a period of great improvements in the roads of Wales, stimulated by the needs of such entrepreneurs as Lord Penrhyn and Thomas Assheton Smith. The Commissioners of the Turnpike Roads in Caernarvonshire became vigilant in enforcing the law that only carts with broad wheels should be allowed special rates. This provoked petitions in 1801 from slate carriers and quarrymen 'at and from the Slate Quarries called Cilgwyn, Hafodlas and Penybryn Quarries in the several parishes of Llandwrog and Llanllyfni in the County of Caernarvon'. The petitioners numbered about 350; and over half of these were small farmers who depended upon the carrying of slate from the quarries to Caernarvon and the Voryd estuary to help towards the payment of their 'Rent and high Taxes'. Their request was granted on the understanding that they would obtain broad wheels; but in July 1801 they petitioned for a further extension of time, pleading that the price of corn had 'continued high beyond all Expectations', so that they had been unable to save any money to buy the broad wheels. They undertook not to beg for any further indulgence, and engaged to work two days each with their carts as directed.[37]

The building of new roads helped the Dinorwic Quarries; but it was not until 1824 that a tramway was constructed across Fachwen to Port Dinorwic, which enabled about 150 tons of slate to be transported in one day. The quay at Port Dinorwic was completed in 1793.

During the time Wright, Ellis and Bridge managed the quarries, Wright acted as estate agent; and John Byrne was employed as agent on the following terms:

> Either to have Sixty Pounds per annum and one two and thirtieth Share in the Adventure or Seventy Pounds per Annum without having any Share in Said Adventure—And to be at liberty to keep two Teams at least to be employed in Carriage of Slate for the said Concern.[38]

On 3 September 1790 William Bridge made an agreement with his son Lynch, which showed that the son was closely concerned with his father in the Dinorwic concern. From 25 March 1789 he was paid £40 yearly 'for his own and his Father's Services'. If the son behaved himself 'in a respectful and dutiful manner to his Father', he was to receive half of the yearly profits of his father's one-third share of the concern, and also half of the profits from the one-eighth share his father had in the brig *Elizabeth* and the sloop *Dinorwic*, both of Caernarvon.[39]

The obstacles in the way of meeting the heavy demand for slate were described by William Bridge to his partner Hugh Ellis in a letter of 26 September 1788:

> If 2,000 Countesses per Day dragged down the Hill from Brynglas and from 1,500 to 2,000 also carr.d from thence on Horse Back will supply the several orders at present for Slate, then you may depend upon being supplied with those Quantities I believe: but as to other sorts or greater Quantities to be sent from the Quarries at present, excepting Tun-Slate, is very uncertain. Every exertion is made and daily continued to be so, to do more, and it will be done if

possible: but Dragging Slate down a steep Hill and carrying the Drag up again is so laborious and difficult a Task, that very few will undertake it at any price.

Eight Memel Dale Balk of 40 to 50 feet long and 13 Inches Square each, are immediately wanted, and I desire you will order them from Liverpool as soon as possible for the Making of a Rail Road to convey the Slates down from Brynglas, to where the Carts can take them up; as we shall thereby with Rope and Windlas, 2 Small Waggons and 3 Labourers, be able to convey all the Slate down from those Quarries as fast as they can be raised and dressed by the Number of Hands now employed at those two Quarries. This appears to be the most eligible way to be put in Practice —both for Expedition and saving Expense. I have put my Engineer upon this Scheme. He has viewed the Premises, measured distances, etc and will be ready to prepare everything as soon as the Balk arrive at Caernarvon. And a good Rope well tarred be ordered to be sent with the Balk from Liverpool of ¾ths of an Inch Diameter and 200 Fathoms in length.

£130 is the amount of Balance due to Workmen and if not paid soon, I must take Flight to avoid being murdered;—Getting Slates carried under your Castle [Caernarvon] will not save me![40]

The long intervals between pay settlements were a common source of irritation to the men, and Thomas Wright found it difficult to collect rents if the quarrymen tenants had not been paid.[41] At Cilgwyn the men often had to wait months before a final settlement was made. The incline mentioned in Ellis's letter had been built by 1789. In that year slates were being sent down to the 'new quay' on the shores of Llyn Padarn; and from there they were taken by boat either to Cwm y Glo or to Pen Llyn. The company had its own boats and crews, and the men were paid 12d per thousand for Countesses, 8d per thousand for Ladies and Patents (ton slates) and 4d per thousand for Doubles. Hired boats were paid at the higher rates of 16d, 9d and 5d.[42] By

1793 there were six carts 'pretty constant' at Allt Wen, and boats were keeping the quay 'quite clear'.

The recruitment of workers was no easy task, and on 9 March 1793 John Byrne wrote to Hugh Ellis:

> I hear there are several more of our Quarrymen intending to go for Lord Penrhyn's quarry at the end of this quarter —It is my opinion that if you or T.W. (Thomas Wright) write to Mr Wyatt on the subject it would save many £100 to both quarries in the year and bring the men to order which is impossible at present owing to the great encouragement they receive at Lord Penrhyn's quarry.[43]

The 'seduction of labour' was common in the early stages of industrialisation, because of the scarcity of skilled workers. Employers usually denied the practice, and undertook to prevent it; and Penrhyn Quarry was no exception. On 9 April 1793 Benjamin Wyatt, Lord Penrhyn's agent, replied to Hugh Ellis as follows:

> Till I received your Letter I had not heard a Syllable of what you have related—I certainly will give directions to Richard Williams to discharge any of the Men who shall make use of such threats—I was at our Quarry last Thursday and found that several of your People had been there to offer to take bargains and had reported amongst the Men that the prices were lower'd—but I did not hear that our Men had anything to say further, perhaps a little reduction which I ordered to be made might afterwards ripen the business, but you may be assured I shall discountenance everything of this sort.[44]

The mobility of labour in the slate-producing districts was considerable, especially among Penrhyn, Dinorwic, Cefn Du and Cilgwyn Quarries. This was partly because wages were so low that even a few pence or a shilling more would be sufficient to encourage men to move. In February 1795 thirty-seven unemployed quarrymen from Penrhyn Quarry, along with their wives and children, were examined at Pentir before being sent back to their 'place of origin'. Thirty-three had come from seventeen different parishes in Caernarvonshire, three from Anglesey and one from Denbighshire.[45] In 1788 there were only slight differences between the list prices for raising slate per thousand at Penrhyn and Dinorwic Quarries:

Penrhyn

Sale price		10s 6d	Doubles	not under	12in × 6in	Raising	4s
,, ,,	£1	1s 0d	Ladies	,, ,,	15in × 8in	,,	8s
,, ,,	£2	2s 0d	Countesses	,, ,,	19in × 10in	,,	16s
,, ,,	£3	10s 0d	Duchesses	,, ,,	24in × 12in	,,	21s [46]

Dinorwic

Sale price		11s 0d	Doubles	Raising	3s 9d
,, ,,	£1	1s 0d	Ladies	,,	7s 6d
,, ,,	£2	2s 0d	Countesses	,,	14s 0d

In February 1789, however, Dinorwic Quarries were setting bargains to at least one quarryman from 'Braich y Cafn' and one from Cefn Du.[47]

No Duchesses or singles (sometimes called 'small doubles') seem to have been produced by Dinorwic at this time, and at Penrhyn only five kinds of tally slates and two kinds of ton slates were produced. By 1830, however, Dinorwic was producing ten kinds of tally slates and three kinds of ton slates. In the lower quarries only ton slates were produced, since the slate rock was 'so close grained and so hard a Nature that very few Countesses and Ladies can be made there'.[48]

Most of the slate was sent to markets in London, Liverpool and Bristol. Freight charges to Liverpool were normally 9s per thousand for Countesses, 5s per thousand for Ladies, 3s per thousand for Doubles, and 5s 6d per ton for ton slates. As slate was liable to breakages, bundles of straw were always placed between the piles on board ship at a small extra charge, from 6d to ½d per thousand depending on the size of the slates.

In 1809 Thomas Assheton Smith took over the management of his quarries in conjunction with William Turner, a Lancashire capitalist, and Hugh Jones, a banker from Dolgellau. Thomas Assheton Smith (1752–1828) had inherited the Vaynol estate from his father Thomas Assheton, who had taken the name of Smith out of respect for his relatives the Smiths of Tedworth in Hampshire. Thomas Smith had acquired the property when it was alienated from the Crown by William III in 1699, although originally the estate had belonged to the Williamses. The estate consisted of the manors of Vaynol and Dinorwic and other estates in various parishes of Caernarvonshire and Anglesey.[49]

On 17 July 1809 a draft case was prepared for the Attorney General's opinion as to Thomas Assheton Smith's right to the quarries in the manor of Dinorwic, as 'about 15 or 16 years ago . . . several persons got into possession of such Quarry, and have continued to work there without the permission of Mr Assheton Smith and now set up an adverse claim'. In 1806 Assheton Smith had obtained an Act to enclose the parish of Llanddeiniolen, which extended to the eastern shore of Llyn Padarn and to the north side of Elidir Fawr; and in 1808 an amendment to the Act was passed, allowing compensation to be given to those who had built cottages on the waste land. The Act described Thomas Assheton Smith as 'Lord of the Manor of Dinorwic within the Parish of Llanddeiniolen and owner of the Soil of the Commons and Waste Lands within the said Parish'.[50] On 10 August 1809 a draft agreement was drawn up by Thomas Assheton Smith, Robert Thomas Carreg, William G. Griffith, David Griffith and John Hughes; and in this agreement Assheton Smith's claim to be lord of the manor of Dinorwic, entitled to all the mines and quarries therein, was acknowledged. To reimburse the men for the large amount of money they had spent on the quarry called Chwarel y Mynydd, Thomas Assheton Smith agreed to allow them to continue working the quarry for four years, paying for three of these four years a rent of one-tenth of the value of the slates. At the end of four years they were 'peaceably and quietly' to deliver up 'the possession of the said Quarry and all Cabins and other conveniences erected upon the Rubbish of the said Quarry' to Assheton Smith.[51]

A similar settlement was made out of court on 11 August 1809 with Glynn Griffith, who had worked Chwarel Fawr; Griffith agreed to work the quarry for four years, not 'to cut any fresh

Ground or Green Sod', and to pay a rent of one-tenth of the value of the slates.[52] In August the following year a quarter share of 'Chwarel Fawr' in the manor of Dinorwic, was put up for auction. It was described as a 'Blue Slate Stone Quarry', and as 'held under a lease, granted by Thomas Assheton Smith, esq., the Lord of the said Manor', which expired 'on the sixth day of September 1813'.[53]

Opposition from the squatters on the common of Llanddeiniolen was more violent than that of these quarrymen. On 20 July 1809 the Commissioners gave notice of their intention to make a survey of the land in the parish of Llanddeiniolen in order to follow out the directions of the proprietors as to how the land was to be allotted and divided under the Enclosure Act of 1808. On 8 September a group of people met upon a 'certain common' in Llanddeiniolen and 'there opposed the Civil Power, and continued together in Riotous and Tumultuous manner for upwards of one hour, after the Riot Act was read . . . then committed several violent Assaults.' Thirteen offenders were named in the *North Wales Gazette* for 21 September, and their descriptions were issued so that they could be identified and brought to justice. Eight of these offenders were quarrymen, and 4 were quarrymen's wives. The editorial comment was that 'these misguided people occupied some cottages, from which they foolishly conceived they could not legally be removed.' The following week the same newspaper reported that all the rioters had been apprehended except three. Praise was given to John Evans, the lawyer and quarry-owner of Caernarvon, for his 'indefatigable and persevering attention in bringing these deluded people to justice, and a due sense of their duty'. The rights of property having been successfully maintained, all the opponents of enclosure were released at the end of October under the comprehensive amnesty associated with George III's Jubilee. Other celebrations at this time included the regaling of Assheton Smith's tenants and the Cilgwyn quarrymen with 'cwrw da' (good beer); and 'Lady Penrhyn, with her usual goodness, distributed two fat oxen to the poor of the parishes of Llandegai and Llanllechid.'[54]

The gross profit of the Dinorwic Quarries from 1 October 1809 to 12 May 1828 was £157,326 2s 0d. From 1 October 1809 to 30 June 1812 the total profit was £5,611 9s 3d; and the annual profits thereafter were as follows:

Year	£	*s*	*d*
1813	3,584	14	3
1814	5,393	14	3
1815	4,931	17	9¼
1816	6,774	18	3½
1817	5,021	19	6¼
1818	8,328	10	7
1819	11,863	12	9
1820	7,307	15	0
1821	6,807	19	10
1822	9,479	19	8¾
1823	10,742	4	7½
1824	13,439	11	8½
1825	16,807	1	7¼
1826	18,030	16	4
1827	14,543	7	11
1828	8,320	3	3¾ [55]

4
NANTLLE AND FFESTINIOG, 1782–1831

Developments in the slate quarries of Nantlle can be traced in the collection of documents known as 'Porth yr Aur', which is preserved at the University College of North Wales in Bangor. A copy of John Wynn's petition for a lease of the mines and quarries in Caernarvonshire, and an abstract of the lease to him dated 25 May 1745, show that he was given the right to dig for slate and minerals in the hills and commons of eight parishes, including Llanwnda, Llanrug and Bettws Bach, for thirty-one years. In January 1774 John Wynn died, and his son Sir Thomas Wynn had his petition for a renewal of the lease refused by the Surveyor General, because John Wynn had 'made up no account of profits during the whole time he had been in possession of the lease'. Sir Thomas Wynn, later Lord Newborough, contended that although hundreds of pounds had been spent on trials, he was never successful enough to work the lease at a profit.[1] It will be recalled that during the period of this lease, quarrymen were helping themselves to slate, and were able to rival Penrhyn Quarry in their sales.

The Crown had been very negligent in the care of its property and land revenue in the early eighteenth century; but in May 1791 Robert Roberts of Caernarvon was appointed superintendent of wasteland and slate quarries. He was responsible for four parishes in John Wynn's lease: namely Llandwrog, Llanwnda, Llanrug and Bettws, and also for Llanllyfni, Clynnog Fawr, Beddgelert, Llanberis, Llanddeiniolen and Llanbeblig.[2]

Roberts proceeded to let bargains, that is, areas of slate, to various companies of quarrymen; and in October, November and

December 1791 and January 1792 he gave sixty-six persons the right to dig for slate for one year from Michaelmas 1791 to Michaelmas 1792, at a rate of 10s 6d each. He received rents from twenty-six people for the 'Liberty of working by themselves or one Substitute for certain quarries in part of a Common or waste called Cefn du (pronounced Kefn dee) . . . and from 40 other persons for the like liberty in part of a Common or waste called Kilgwyn'. At this point the Agent of the Crown was interrupted in his duty by the return of Lord Newborough, who immediately set up a claim to the quarries in Cilgwyn; and Newborough's agent Samuel Price ordered the workmen not to pay any more rent, at the same time circulating a report that the Crown had no right to the quarries in Cefn Du. As a consequence all the quarrymen, except four, refused to pay any more rent. The four men who paid, on condition they would be protected in possession of their quarry, were attacked by a mob. The attackers threw rubbish in their way, and blocked a road which they had made. Although they were warned by Roberts, they continued to obstruct the four quarrymen; and finally they took over the working of the quarry themselves. Roberts had then no choice but to repay the quarrymen's money, since he was unable to afford them protection.

In March 1795 the case of the men who wanted their money back was considered: the sum involved was £34 13s. The situation was a delicate one: although John Wynn's lease had expired in 1776, Lord Newborough's agent had continued to pay, and the Crown to accept, the old rent of 10s per annum, until 1790, when it was still offered, but not accepted. Lord Newborough had been refused a lease of the same large extent as formerly, although he had been offered one of more limited scope, which he had declined; and no formal notice had ever been given to Lord Newborough to give up his quarries. In April 1794 twenty actions of trespass were brought against Roberts at Conway Assizes, relating to the quarries in part of the mountain called Cefn Du, a tract of common land running into several parishes. The proof of the Crown's right to the land, mines and quarries rested on 'general reputation and . . . Acts of ownership and possession exercised by the Crown and the Grantees'. The Crown's right was admitted by Sir Thomas Wynn's petition in 1773, in which he asked for a new lease including Llanbeblig where slate quarries had been recently worked under a bargain from Roberts. Leases had also been granted in 1757 and 1764 to work the mines in the parishes which included Cefn Du. These mines had not proved valuable, and few were 'now wrought'; but on all the leases rents had been paid to the Crown.

The Crown, however, was well aware of the dangers of bringing an action to court, since Lord Newborough's agent was taking all measures to make the defence of the Crown's rights extremely unpopular. He was persuading the 'County Gentlemen that it is essential to the Preservation of their own Rights, to resist those of the Crown on the present Occasion; and though such an Insinuation is utterly groundless there is Reason to apprehend that he will find it no difficult task to influence the Country ag.t the Business, as most of the landholders have made Incroachments on the Waste Lands and other Rights of the Crown, and the Quarrymen naturally prefer holding under Lord Newborough, who has hitherto required from them an annual acknowledgement of only four pence per Man which acknowledgement Mr Price says has never exceeded 14s in any one year and that he says has uniformly been expended by Lord Newborough in an Annual Dinner to the quarrymen thus making use of this property of the Crown for the sole purpose of increasing his own Influence in the Country; whereas the Agent appointed by your Lordships is not authorised to let any Quarries for less than 5s per Man and has let none hitherto under 10s 6d which is indeed the common price required by private owners. It will not therefore be surprising if the minds of a Jury of that County be found not quite free from prejudice when this Business comes before them.'[3]

In April 1800 the matter was partly settled, largely because of the efforts of John Evans's friend W. Harrison of the Land Revenue Office; a lease of Cilgwyn Quarry was granted to John Evans, Richard Roberts, a merchant, both of Caernarvon, John Price of Anglesey and Thomas Jones of Bryntirion in Caernarvonshire. There was an earlier attempt by Lord Newborough, in April 1798, to petition for a new lease; but he decided in the following year not to oppose John Evans.[4]

John Evans entered into articles, in August 1789, with Hugh Ellis, the second son of Archdeacon Ellis, vicar of Bangor. Hugh Ellis and Thomas Wright had been working the Dinorwic Quarries since 1787, and in 1788 John Evans was working as a clerk on the Vaynol estate. It was there that he gained his experience in the slate business, so it was natural that he should play a leading part in the development of Cilgwyn and Cefn Du Quarries. John Evans took over Hugh Ellis's practice completely when he died in 1808; and he extended it considerably. He administered several local estates, and he earned a reputation in slate quarrying so that he was consulted over administration and shipping as well as litigation. In many cases he seems to have been suspected, probably wrongly, of sharp practice. That his company was often unscrupulous in pursuing its own ends was apparent in its policy towards Cefn Du. This common and its quarries were not included in the original lease for Cilgwyn, although in December 1799 a vein of blue slate had been discovered there by 'a poor man' who was building a cottage on the common. The company initially bought a share of the quarry in Richard Roberts's name; and, under his advice, it was to be worked 'quite wrong so that but a few Slates will be got out of it'.[5] This was planned as a means of distracting attention from a potentially valuable quarry; but as quarrymen were damaging the old quarries of red slate on Cefn Du common—'in fact they completely Gut them of slate and fill them with rubbish'—the company applied for a lease of this common.[6] It was granted, after some delay, in 1803.[7]

Financial arrangements were made on 16 May 1800, whereby the four partners agreed to open an account with the North Wales Bank at Caernarvon under the name of 'the Cilgwyn and Cefndu Slate Company' (at this date, they rented the quarries of red slate on Cefn Du common), and to deposit £100 each towards carrying on the activities of these quarries.[8] Cilgwyn covered an area of 150 acres and was bounded on the south and south-west by enclosed land, containing slate quarries including those of Talysarn and Hafodlas. Cefn Du was approximately 300 acres and was bounded on the north-east by the Llanberis lakes and on the south-east by Glyn.[9]

Difficulties from workmen who refused to accept the validity of the Crown's lease of Cilgwyn quickly developed. In July 1800 John Evans went to London to discuss this question and that of the lease of Cefn Du. He had a successful interview with Harrison, but London appeared like 'Babylon' to him, and he was eager to return to 'the Mountains of Wales'.[10] The following notices about twelve trespassers were distributed by John Evans on 15 September 1800:

> Whereas the Right Honorable the Lords Commissioners of His Majesty's Treasury have granted a lease of Cilgwyn Common and Quarries to John Price, Thomas Jones, Richard Roberts and myself which lease commenced on 5 April last. I do therefore give you notice of such Lease and I hereby require you not to enter into or upon the said Quarries or any of them . . . or in any wise work therein otherwise you will be proceeded against.[11]

These notices had little effect, and on 25 November 1800 Richard Roberts wrote to John Evans to complain of the intruders, described by him as 'Indians', saying that he had been 'afraid of a Bloody war', but that nothing yet had come of it.[12] The quarrymen's champion, on this issue, was Lord Newborough, who took up their case, and supported them against the new company. On 1 December 1800 the four partners wrote to Whitehall, describing the situation and endeavouring to prove that Lord Newborough was 'the greatest enemy to the poor in Wales by prevailing upon these Men to oppose us'.

The lease granted to the partners was for 31 years with an annual payment of one-tenth of the value of the slate they quarried; and when they took over the quarries in April 1800 they found 3 groups, 32 men altogether, working the scattered quarries. The partners employed about 30 to 50 'poor Labourers residing in the Neighbourhood who were not able to find any employment and consequent of the high price of Corn were with their Families reduced to the greatest distress, Slate being the only article for Export from the county of Caernarvon'. The partners' arguments were sound. The Napoleonic Wars had created unemployment and distress, and the harvests of the years 1800 and 1801 were disastrous. Foreign corn could not easily be imported; and the consequent high prices, with wheat over 100s a quarter, meant near starvation for the poor.

The partners went on to say that at first they attempted to co-operate with the trespassers, requesting them to pay one-tenth of the slates as rent. This they refused to pay, as Lord Newborough had informed them that the Crown had no right to the quarries, and that he would defend them if legal proceedings were taken against them. As the work in the quarries at this stage was that of driving levels and clearing rubbish, left by the 'poor Labourers' who because of poverty had worked them improperly, the company could only employ a small number of quarrymen, and others had been obliged to go to 'Lord P——'s Quarry about fifteen miles from their Families'.

The company intended to 'build a large warehouse near the Quarry in order to keep a proper Stock of Corn to supply our poor workmen at as moderate a price as will be possible'. The charge made by Lord Newborough about the ill-treatment of the independent quarrymen was refuted, and attributed to his resentment over the non-renewal of his lease. The concern was described as 'one of the most laudable and best undertakings for the welfare of the Labouring poor that could take place', particularly as employment was then 'so difficult to be got and corn so dear'.[13]

It proved almost impossible to overcome the independent quarrymen and their supporter Lord Newborough, and the efficient working of the company was thereby seriously impeded. In February 1802 it was reported in the statement of the case that:

The Quarrymen of Cilgwyn in consequence of being supported by Lord Newborough became riotous and declared in the most violent manner that they would not quit, and in fact assaulted the Workmen of the intended new Lessees and prevented them from working except in such parts of the Quarry as they thought fit, other Quarrymen have in like manner possessed themselves by force of the greatest part of the Quarries on Cefn, and they continue to do so to this time in the very great injury of the Crown and the new Lessees. They deny that the Crown has any right to the Quarries, but assert that they belong to Lord Newborough, and threaten to kill the first Man that attempts on behalf of the Crown or the new Lessees to dispossess them—and whenever they are applied to, to desist, they sound a Horn and collect in a Body of from 60 to 100, and declare in the most threatening manner that they would lose their lives sooner than quit possession of the Quarries.

There are from 120 to 200 Intruders upon Kefndu Common and from 60 to 80 upon Kilgwyn. From their great numbers and violent dispositions, there is no chance of getting them off by persuasion or remonstrance, therefore it is now become absolutely necessary to take immediate steps to dispossess them, otherwise the right of the Crown to the Extensive Waste Lands in the principality of Wales will be entirely lost, which is of great importance to the Crown from the great Treasures in Mines and Quarries they evidently contain.[14]

Notices were again served on the trespassers in April 1802, warning them that they would be proceeded against 'in his Majesty's Court of Exchequer, at Westminster' if they persisted in their attempts to quarry slates.[15] By 1804 there were only 10 quarrymen at Cilgwyn who were defying the lessees; but 'from the late Great Demand for Slates' there were from '150 to 200 Intruders upon Cefndu Common'. These men were quarrying slates rent free and could undersell the lessees.[16]

These severe handicaps to the company's progress were reflected in the accounts for the quarries. For the first year 'total expenditure at the Cilgwyn Slate Quarry since the Commencement of the Work to the Conclusion of Spring Quarter, 1801' was:

To Workmen's Wages and Sundries	£627 11s 6d
To Expenditure of the present quarter	67 15s 2d
	£695 6s 8d
By slates made £3 12s 6d	

This was a poor return; but at Cefn Du production was more promising. For the same period total expenditure was £418 3s 9¾d, and the output of slates consisted of: '56 tons 10½ cwt Ton Slate, 34,500 countesses, 113,100 ladies, 166,150 doubles, 65,100 small doubles, and 10,500 singles'. The value of these slates was £282 11s 4d.[17] By 1 September 1803 the company had a balance of £815 10s 3d, a profit of £77 3s 11d having been recorded for that year. At the end of 1803 the company appointed Richard Roberts junior, the son of one of the partners, to be their agent in London; he was to take contracts for 'any Quantity of Welsh Slates' from Cilgwyn and Cefn Du Quarries, the orders being sent to the company at Caernarvon.[18]

In August 1804, notice was given of an intended application at the next session of Parliament for leave to bring in a bill for enclosing certain waste lands in the parishes of Llanllyfni, Llandwrog and Llanwnda, all in Caernarvonshire. These commons or waste lands were named as 'Clogwyn Melyn, Cilgwyn, Kini, Moeltryfan, Rhos y Gadfa, Gallt y Coed Mawr and Braich Rhydd'; and most of them contained slate quarries. This represented another attempt on the part of the company to strengthen their position and get rid of trespassers. W. Harrison, from the Land Revenue Office, Whitehall, wrote on 23 February 1805 to John Evans, saying that he had put the matter before the Surveyor General. The latter was not in favour of enclosure, as it concerned 'a question of *right* (the Parties working the quarries claiming a right so to do)', and the likelihood of Parliament passing an Act in that case was slight. He also believed that even if the Act were passed it might remain, like the enclosure of 'Rhos yr Waun' unexecuted. (On the Crown common of Rhoshirwaun in Lleyn, fishermen had resisted the execution of the enclosure Act passed in 1802; and the final award was not made till 1814.)[19]

The Surveyor General considered that the 'Rhos yr Waun' Act had not been executed because of the magistrates' 'lack of vigour and energy, or perhaps because of a mistaken notion of humanity'. He urged that actions against the trespassers should be tried in 'one of the neighbouring English Counties': 'the making Examples of a few, by Imprisonment or levying of Costs, would be the most likely means of restraining others.' Another remedy he suggested was to bring an action for combination or conspiracy against the labourers and 'opulent persons' who were opposing the rights of the Crown. These measures were necessary, as the 'Intruders perhaps conceive that the Lessees are intimidated, and derive courage from their supposed fears.' The Surveyor General said that in 'all questions of right' the common law of the land was 'the proper resort'.[20]

This advice was speedily acted upon, and the threat of legal action resulted in a declaration, on 21 March 1805, by 4 men 'being Trespassers and Intruders Quarrying and working Slates upon a certain Common called Cilgwyn', that they would not in future 'disturb or Molest' John Price (the son of the original partner who had died), Thomas Jones, John Evans and Richard Roberts 'in the Possession of the said Waste Lands or in working any Quarries'.[21] At the same date Captain Robert Evans of Caernarvon likewise gave up his possession of the quarry called 'Vaingoch' at Cilgwyn; and in return for this, proceedings against him were stopped.[22]

For several years 'the old quarrymen' gave up working the quarry known as 'Vaingoch' or 'Vaincoch' on Cilgwyn Common; and the company's railroad above this quarry brought slate waste very close to it. In 1810 the railroad was altered and waste from the wagons was emptied nearer the old quarry. 'The old quarrymen' then came up to the railroad and threw several of the wagons over 'a very high Precipice'. An action was brought by the company to recover damages. The defendants pleaded their 'absolute right' to this part of Cilgwyn Quarry, maintaining that they and their fathers had been in possession of it for forty years, notwithstanding that they had acknowledged Lord Newborough's lease from the Crown.[23] The final outcome of the case is uncertain; but, after receiving notices of the action the 'Oppositionists . . . behaved in a most riotous manner and threatened to kill any person that would come near them, and then broke another Car when another action was commenced'. The total costs of the case were £22 13s 10d.[24]

Trouble of this nature continued for many years: even as late as 1834 a case of trespass occurred.[25] This was not surprising, as the commons of Cilgwyn and Cefn Du contained numerous small quarries which had been worked locally from time immemorial. Many of these quarries were taken over by the Cilgwyn Company and then sub-let. In December 1812, for example, Thomas Bulgin a Bangor slate merchant, John Pritchard a yeoman quarryman, Rice Price a Bristol slate merchant and John Parry a Chester stationer took a lease from the Cilgwyn Company of Cook's Quarry on Cefn Du Common, for fifteen years at an annual rent of 5s and a royalty of one-ninth of the value of the slates made.[26]

The Cilgwyn Company wished to buy Hafodlas Slate Quarry; and they were offered it and the farm for £4,000 by the Rev R. M. Humphreys in 1800. They refused to buy at this price; and for a time Humphreys worked the quarry himself. In July 1812, however, he entered into a partnership with Richard Thornton a merchant of Liverpool, Ellis Williams a quarryman of Llanllyfni and George Bettiss an innkeeper of Caernarvon.[27] Moeltryfan Quarry, included in the Cilgwyn lease of 5 January 1825 which gave the company an additional portion of the common, was sub-let in 1823, before they had authority to grant a lease.[28] The termination of the agreement with William Jones a joiner and John Jones a shoemaker was the subject of litigation, which continued from 1827 to 1830.[29]

In February 1804 John Evans obtained a lease of slate quarries upon part of the Coetmor estate near Bethesda, which included the Ty'n Ffridd Quarry, from 'the Rev Mr Wynn'.[30] This was sub-let in 1812;[31] and in July 1825 John Evans obtained a tack note from J. Griffith, the owner of the land, for a lease to make trial for slates upon Bryn Hafod y Wern.[32] Manod Quarry, on the Manod and Moelwyn Commons in Merionethshire, was leased on 29 April 1805 by James King of the Office of Woods and Forests to Owen Anthony Poole (who by January 1807 had become one of the partners of the Cilgwyn Company) and John Richards, for thirty-one years subject to a yearly rent of 20s to be paid to King plus one-tenth part of slates. Poole and Richards gave up their trials for slates in 1813; and they let the quarry to Francis Webster, an architect, and 5 others, who agreed to be tenants or 'adventurers' of the quarries that might be found on Manod Common, an area of 500 acres, for twenty years from 1 July 1813. It was a partnership of 8 shares, and John Evans was to be one of the partners; but his name, 'for reasons known to the said parties', was not to be introduced into the lease. Thomas Bulgin was one of the partners, although in June 1813 he thought it a doubtful proposition 'and the situation so high that it can only be worked about four months in the year . . . and owing to the very Steep Road to it, it will be very difficult to get the Slates carried down'. He decided to take the quarry, hoping to persuade 'a few of my English Friends to join me in this uncertain speculation'. This he succeeded in doing; but by 1815 another of the partners, Joseph Robinson, reported to John Evans that the level had turned out badly and there was 'no hope of finding a quarry on Manod Common'.[33]

Bwlch Slaters, at the head of the Machno Valley, a slate quarry which was part of the Manod Quarries, was leased by John Evans to John Pritchard for seven years in May 1824; and in July 1824 a merchant called James Smart took over three-quarters of the quarry from Pritchard.[34]

Gallt y Llan Quarry in the parish of Llanberis was leased by Richard Roberts, one of the partners in the Cilgwyn Company, to Owen Parry, a quarryman, and Hugh Hughes, a slate loader, in October 1811 for ten years, at a rent rising from £20 for the first year to £30 annually from the third year onwards.[35] Richard Garnons, the owner of the estate called Penybryn and Talysarn in the parish of Llanllyfni, leased the slate quarries to Hugh Jones, William Turner and William Wynne in September 1801 for a term of thirty years, with one-eighth of the value of the slates to be paid to Richard Garnons.[36]

The most profitable concern of all the many quarries organised by John Evans and his partners was Cilgwyn Quarry; but even this only produced, from 20 January 1807 to 31 January 1812, a net profit of £2,614. After 1812 most of the other quarries were sub-let on short leases; and in that year Richard Roberts ceased to be a partner, and Edmund Francis became agent to the company having 'the sole management of the concern'. He continued in this position until 1827. O. A. Poole died in February 1823 and Thomas Jones in November 1824; and in July 1827 John Evans died. His nephew Evan Evans succeeded him, not his son who had been estranged from his father by an unfortunate marriage. Evan Evans was not enthusiastic about the slate concerns of his uncle, and in February 1829 he wrote to John Price that he was 'altogether tired of Slate Quarries'. He went on to make the suggestion: 'What say you to advertizing the sale of the whole Concern in London for in this Country I fear we are too poor.' In 1830 the partnership was legally dissolved.[37]

The quarries in Merionethshire at first merely provided the local inhabitants with slate rock from which they made slates for their own dwellings. Farmers had small quarries on their land, and one of the earliest records of quarrymen at Ffestiniog was in the registers of the parish church in 1772 and in 1785. The entries showed that 3s 6d was paid to a number of quarrymen for raising two loads of slate to roof the parish church. The slate was extracted from an old pit at Cefn Bodlosgad, the site of an early farm. These slates were of no standard size.[38]

Writing at the beginning of the nineteenth century, Edward Pugh found Ffestiniog 'a small poor village'; but he added that 'the great slate quarries' belonging to Casson & Company were in 'a romantic spot'. The slates, 'fine blue stone and of large sizes', were exported to Dublin and other places, but their conveyance to a quay a mile beyond Maentwrog was 'both tedious and expensive: these gentlemen not being, as yet, possessed of a railroad, they employ for this purpose about thirty carts, which, certainly do not *improve* this public road'.[39] This quay was on the River Dwyryd, and the slates were sent from there by barge to the coast. At first there had been no road from Diphwys Quarry to the main road; the early workers had carried slates on their shoulders till they reached the road, and then sold the slates in the neighbourhood.[40] The narrow-gauge railway from Ffestiniog to Portmadoc was not built until 1836.

The Ffestiniog slate quarries were situated on the sides of a hollow between the Manod and Moelwyn Mountains; and the most valuable slate was found in the centre of the region, near Blaenau Ffestiniog. The slates, which occur in the Llandeilo Series of the Ordovician System, are mainly blue or grey in colour, and are very fine-grained. The rocks lie east to west with a dip due north and north-west. The slate beds are between bands of hard sandy rock, and a bed of good slate is known as 'Llygad' or 'eye'. There are about seven beds or veins, including 'Old Vein', which was the first to be worked, 'New Vein', and 'Back Vein'. Open quarrying is not an economic proposition, as it would mean that a large amount of useless rock would first have to be removed because of the strata's low inclination (less than 45 degrees) to the horizontal. Mining was thus used near Blaenau Ffestiniog, at Corris, at Aberllefenni and in Denbighshire. In South Merionethshire and the adjacent part of Montgomeryshire there are two chief veins in the slate belt: the 'Broad Vein' and the 'Narrow Vein'. In the Corris district the Narrow Vein is the most important band. Between the Broad and Narrow Veins there is the Middle or Red Vein, where the slates have a tint of rust.[41]

In Merionethshire, to a larger extent than in Caernarvonshire, English capital made a great contribution to the development of the slate industry; and men such as the Hollands, the Cassons, William Turner and the Greaves family played a memorable part in the industry. The first quarry to be worked in Blaenau Ffestiniog was at Diphwys, where in 1765 Methusalem Jones and William Morris, two quarrymen from Cilgwyn, successfully established a quarry. The working was started on Lord Newborough's land; and in 1794 eight quarrymen formed a new company, paying an annual rent of £120 per annum for the quarry and £20 per annum for a farm near the quarry. The workers only numbered about twenty, and the company had very small resources. About 1799 William Turner asked Lord Newborough for a lease of the quarry, and this was granted.[42]

William Turner was born at Seathwaite in Lancashire in February 1766, and his interest in slate derived from the fact that his father was the lessor of the Walmascar slate quarries in that district. In order to purchase the Diphwys Quarry, Turner sought help from his friends Thomas and William Casson of Cumberland; and the quarry was run by the three men in conjunction. They needed more money to conduct it properly; and so Hugh Jones the Dolgellau banker became a partner, and the concern was called first 'William Turner & Company' and later 'Diphwys Casson Slate Company Ltd'. The firm obtained wartime government contracts, and sent slates to cover the barracks at Portsmouth, Plymouth, Dublin and Cork. They made a road from Diphwys to the parish road, and large quantities of slates were carted to Tanybwlch; from there they were taken in barges to Ynys Cyngar, an island opposite to what is now Portmadoc, where the barges discharged into small brigs which carried the slates to the place of sale.

It is probable that a considerable fleet of small brigs was employed by the company, as Sir Llewelyn Turner found a vast number of bills of lading among his father's papers. A typical bill of lading was as follows:

> Shipped, by the Grace of God, in good Order and well Condition'd, by William Turner & Company in and upon the good ship called the *John*, whereof is Master, under God, for this present Voyage, Lewis Lewis, and now riding at anchor in the Port of Traeth, and by God's Grace bound for London with slates, to say, 9,500 Dutchesses, 11,000 Countesses, 9,000 Ladies, 18 Tons Queens and 17 Tons Rags, being mark'd and number'd as in the Margin, and are to be delivered in the like good Order and well Condition'd,

at the aforesaid Port of London (all and every the Dangers and Accidents of the Seas and of Navigation, of whatever Nature and Kind soever, excepted) unto Mr Owen Hughes or to his Assigns, he or they paying Freight for the said Goods according to agreement with Primage and Average accustomed. In Witness whereof, the Master or Purser of the said Ship hath affixed to three Bills of Lading, all of this Tenor and Date; the one of which three Bills being accomplished, the other two to stand void. And so God send the Good Ship to her desired Port in safety. Amen.

Dated in Traeth, April 9, 1811[43]

William Turner, Hugh Jones, William Casson and Thomas Casson regarded themselves in 1823 as 'Slate Merchants carrying on Business at Ffestiniog in N. Wales under the firm of William Turner & Company.' This had been the case 'since 1806'; but 'the active part of such Business' had always been carried on by William Turner 'alone on behalf of himself and his partners Hugh Jones, William Casson and Thomas Casson'.[44] The company's accounts were kept with the firm of Thomas and Hugh Jones in Dolgellau, and the accounts from May 1813 to July 1814 show that at that date large profits were being gained:

July 19 1813 Cash transferred into Mr H. Jones' Account being his proportion of the profit to the 8th of May last as per Account stated . . . £4,108 6s 5d

the like proportion transferred into Mr Turner's Account . . . £4,108 6s 5d.[45]

William Turner and Hugh Jones joined Thomas Wright and Thomas Assheton Smith in October 1809 as partners in the Dinorwic Slate Quarry. Assheton Smith had a half share, Thomas Wright a quarter share, and Jones and Turner an eighth share each. The partnership lasted until 1828, when Assheton Smith died. The total amount paid to Turner and Jones 'for Gain' from 1 October 1809 to 12 May 1828 was £24,843 7s 10¼d. On 28 September 1820 Thomas Wright died; and his share of the quarry was taken over by Assheton Smith. In 1824 William Turner was superintending the quarries, and his payment for this from 30 September 1825 to 30 June 1826 was £2,253 16s 11d.[46]

The list of quarries worked at one time by William Turner was very long. In addition to Diphwys and Dinorwic they included Coetmor at Bethesda and Penyrorsedd, Penybryn, Cloddfa'r Coed and Dorothea, all of which were in Nantlle. Turner died on 7 November 1853; and on the day of his funeral the inhabitants of Caernarvon closed the shops in token of respect for 'this venerable gentleman', who, according to his son, 'had never been guilty of any base, indirect or sordid arts'.[47]

Another pioneer of the Ffestiniog slate industry was Samuel Holland junior, who was born in Liverpool on 17 October 1803. His father had been in Wicklow with William Turner looking for gold, silver and lead; but when the rebellion broke out in 1798 they had left Ireland and come to Wales. Turner began to work the Diphwys Quarry, and Samuel Holland senior became a speculator in Welsh quarries and mines. One of his first speculations involved Lord Penrhyn: Holland and his partner, Michael Humble of the firm Humble & Holland which owned the Herculanean Pottery Works at Loxteth Park, undertook the sale of slates from Penrhyn Quarry. They summoned Samuel Worthington from the pottery works to act as their agent at Bangor for the sale of slates; and although Humble and Holland ended their connection with Lord Penrhyn in 1811 or 1812, Samuel Worthington continued to work for him.[48]

An agreement between Lord Penrhyn on the one hand and Worthington, Humble and Holland on the other was made on 1 August 1805. This said that the three men were to be supplied with 'all the slates made and manufactured at the quarries of Cae Braich y Cafn'; but special provisions were made for John Furey of Winnington; Samuel and James Wyatt of London; Barker Chifney of London; and William Hayes of Frodsham in Cheshire, 'he being an old Customer'. The price of tally slates at '1,260 in Tale on the waggon for 1,000' were as follows: Duchesses 70s per 1,000; Countesses 40s per 1,000; Ladies 20s per 1,000; Doubles 11s per 1,000; and Singles 5s per 1,000. A price of 18s per ton was charged for 'Rags of Ton Slates 21 cwt to the ton when on the waggon at weighing and reckoning 6 score pounds to the cwt'. These prices were the same as those being charged in 1798.

The weight of the tally slates was reckoned according to the following table:

Every	1,000	duchesses	at 3	tons
,,	,,	countesses	at 2	tons
,,	,,	ladies	at $1\frac{1}{4}$	tons
,,	,,	doubles	at $\frac{3}{4}$	tons
,,	,,	singles	at $\frac{1}{2}$	tons

Holland, Worthington and Humble, together with their 'agents and servants' were to occupy 'the new Rooms at the Quay now occupied by William Williams and Robert Thomas and also . . . a proportion of the wharf or quay not exceeding one half thereof as shall be necessary to the lying down and shipping of the slates.' The slates were to be paid for by 'a good bill on London'.[49]

After he had ended his agreement with Lord Penrhyn, Samuel Holland worked a quarry on Cefn Du Common; but in 1819 he heard about a promising quarry at Rhiwbryfdir near Ffestiniog, on land owned by William Oakeley of Tanybwlch. The quarry proved to be 'a hole' with three men working in it; but as these men had no lease, Holland acquired in 1820 a 'take note' which gave him permission to work there for three years. In March 1821 Samuel Holland senior asked his son, who was in Liverpool, to go to Rhiwbryfdir and take over the slate quarry. Samuel Holland junior found the journey a test of endurance and intelligence: as there was no road from Penmachno to Ffestiniog he had to go over the mountain by cart road. He managed to find 'Holland's Quarry', which was still only a small hole; and his father then gave him his instructions before leaving him in charge. At this date Samuel Holland junior was only eighteen years old.

Farming and quarrying were almost of equal importance to Samuel Holland junior, and he kept from 1,500 to 2,000 sheep on the mountain around Rhiwbryfdir Quarry. Drovers used to come and take the sheep off to English fairs, Holland selling them at 12s to 14s a head.[50] The slate business flourished also: from 1824 to 1825 the royalty paid to William Oakeley, at one-tenth the value of the slates produced, was £25,000.[51]

In 1825 it was rumoured that a new company, the Royal Cambrian Company, was being formed in London for the purpose of working mines and slate quarries on Crown lands in Wales; its capital was said to be around £500,000.[52] The nobility and gentry were alerted to this 'projected Company of Strangers'; but in November 1825 Nathan Meyer Rothschild, a leading promoter of the Royal Cambrian Company, gave notice of an application to Parliament for leave to bring in a Bill for making a railway from the New Moelwyn Quarry to Portmadoc.[53] In the previous month W. A. Madocks, the builder of the embankment at Traeth Mawr and of the harbour at Portmadoc, had attempted to bring in a Bill for a railway to Portmadoc from the slate quarries of Lord Newborough, W. G. Oakeley, S. Holland and W. Turner & Company.[54] Thanks to the rivalry between the two Bills and to opposition from landowners and carriers, neither Bill was successful and the project was delayed until May 1832, when despite continued opposition Madocks's Bill (now taken over by Samuel Holland) was passed.

Armed with a lease from John Wilkin, the Receiver of the King's rents for North Wales, Rothschild attempted to search for slates and other minerals on lands and sheepwalks adjoining Rhiwbryfdir Quarry. But he was successfully challenged by William Oakeley, who served notices of trespass. The matter was eventually settled out of court, and Rothschild paid costs and damages for trespass. The case, which was regarded as a victory for the landowners against the Crown, 'excited no small degree of interest among the gentlemen possessed of Upland estates in the principality of Wales'.[55]

The years 1824–5 were notorious for speculation: boom conditions prevailed until 1826, when a slump set in. Nathan Meyer Rothschild, known in Wales as 'the great speculating adventurer in mines, quarries etc', formed in London a new company called the Welsh Slate & Copper Mining Company. It had Lord

Page 85 (*above*) Votty and Bowydd Quarry, showing a slate block being drilled for pillaring by a hand-operated jumper; (*below*) Votty and Bowydd Quarry, showing holes being drilled for channelling in an underground chamber lit by candles

Page 86 (*above*) Votty and Bowydd Quarry, with slate blocks being lifted by hand-worked pulleys; (*below*) Votty and Bowydd Quarry, showing slate being split with a mallet and chisel

Palmerston as its chairman; and unlike the New Moelwyn Company, which was short-lived, it was ultimately very successful. In February 1825 some of the directors came to Wales and inspected Rhiwbryfdir. By this date there was a new road; and Holland had persuaded some of the farmers to carry slates in wagons to riverside wharves at Cemlyn, Gellingrin and Cei Newydd below Maentwrog. From there they were taken by barges to the harbour west of Traeth Mawr, which had been made by W. A. Madocks in 1804 and was called Portmadoc.

The Welsh Slate & Copper Mining Company had decided to concentrate on slate rather than copper; and they were anxious to acquire an interest in Rhiwbryfdir Quarry from the Hollands. After much negotiation the parties agreed on prices of £25,000 for the quarry and £3,000 for the farm. Samuel Holland senior retained the land above Rhiwbryfdir Quarry; and he worked the quarry there, which was known as Holland's Quarry and in 1828 was handed over to his son. The son ran into legal difficulties with the Welsh Slate Company when he drove a tunnel or level through land which they claimed as theirs. The ensuing legal battle was won by Samuel Holland junior, but it was protracted until 1833. Shortly afterwards Samuel Holland was granted an extension of his lease, and the royalty was reduced; he continued to work his quarry until 1877.[56]

Lord Newborough's quarry in Merionethshire was begun about the year 1801. It was later called Bowydd Quarry, but for many years it was known as the Lord Quarry. It adjoined Diphwys Quarry but Lord Newborough invested little capital in the concern, doing 'little more than dig a large hole'. Even this was given up; and the quarry remained unworked until 1823, when Lord Newborough's brother re-commenced operations. In 1827 the quarry was taken over for a short time by John Roberts of Caernarvon; and in 1834 John Greaves and Edward Shelton acquired the quarry and began to develop it more successfully.[57]

5
ECONOMIC, SOCIAL AND TECHNICAL DEVELOPMENTS, 1782–1831

The output of the slate quarries in Caernarvonshire increased from under 20,000 tons in 1786 to over 90,000 tons in 1831; and that of Merionethshire increased from about 500 tons to over 12,000 tons in the same period.[1] The expansion of the industry was largely due to the growth of population, and to the Industrial Revolution which produced an urgent demand for roofing materials for houses and factories. The demand was by no means a steady one; and the slate industry, like brick-making and building, underwent periods of activity followed by stagnation. During the period 1731–82 the industry had been on a very small scale, with the Nantlle quarries successfully competing with Penrhyn Quarry; but from 1782 to 1793 there was great interest and development.

The war-time depression lasted from 1793 to 1799; but for the rest of the war trade was good. The end of the war in 1815 brought back the depression, with the prices for slate falling; but from 1816 to 1819 the building trade improved, and there was a short-lived boom in 1819. This was followed by a slack period until 1822, when trade again revived to reach a peak in 1825. This was followed by stagnation and a fall in prices of nearly 30 per cent.[2]

The early progress of the slate industry was checked very seriously by the outbreak of war with France in February 1793. The industry reached an estimated peak production of 26,000 tons in 1793; but the output of Penrhyn Quarry alone fell from 15,000 tons in 1794 to 8,000 tons in 1796. The cost of sending slate to London by water rose from 12s 9d per ton in 1790 to 36s in 1794 and 38s in 1814.[3] Insurance rates for cargoes on the London route, where the risks were greatest, rose steeply; and within weeks of the declaration of war the Dinorwic agent John Byrne was writing to Hugh Ellis:

> This great talk of War has stopped all the London Vessels from proceeding to their voyages. The last two that I loaded for Mr Morris are here yet—they all talk of now sailing as far as Millford and wait there for a Convoy. Sevl mean to try the Bristol Channel for to sell their slate, in order to save their going round the land. There is not one Vessel that will take a cargo now for London. The Insurance run now to £10 per cent which is more than the freight—I think it lucky to have the Runcorn Trade at present.[4]

The Napoleonic Wars hindered building and this in turn checked the slate industry; but one of the worst and longest-lasting consequences of the war was the heavy tax which was levied on slate. A general tax was in a sense overdue: since 1784 bricks and tiles had been liable for duty, and in February 1794 the government imposed an additional duty of 1s 10d on every thousand plain tiles made in Great Britain. The total tax on such tiles thus became 4s 10d; but in the same month it was resolved that a duty of 10s should be charged on every ton of slate, produced in Great Britain, which should be carried coastwise.[5]

The news of the tax roused the people of Caernarvonshire to try to secure its repeal or modification. The tax was highly discriminatory, as it did not affect quarries situated near navigable rivers and canals, being levied only on slates carried coastwise. Penrhyn and Dinorwic Quarries were affected very badly by the tax; and Benjamin Wyatt, the Penrhyn agent, wrote to Hugh Ellis on 26 February 1794 supporting plans for a meeting of the county. Wyatt said that one of their Staffordshire customers had already written to him, saying that when the tax came into operation he would have to cease buying from them and buy slates instead from the Swithland Quarries in Leicestershire. These quarries employed 150 men, and would soon be linked by branch canal to the Leicester Canal: they produced thick green and purple slates, which were ultimately to give way to the thinner Welsh slates. Wyatt pointed out that tiles were generally made on or near the spot where they were used, and that exclusive of carriage the value of slates and tiles was nearly the same:[6] he did not foresee that with the development of canals and railways the local monopolies of many tile manufacturers would be broken. In 1794 Penrhyn and Dinorwic Quarries were at a disadvantage as compared with the Lake District slate quarries, which could send their slates by canal.

Wyatt sent Ellis a copy of the 'Humble Petition of the Slate Merchants, Slaters and Builders in Birmingham and its Vicinity', which was to be presented to the House of Commons. This affirmed the petitioners' acceptance of the need for a tax on slate; but it objected to the proposed tax which would give tiles 'such a decided superiority over Slate that the latter which gives bread to Thousands and is an article of the first importance to Lancashire and North Wales will be thrown out of request and ruined'. The petitioners asked for a less partial tax, 'just in its nature and more productive'. Wyatt added that if they failed to get the duty abolished they should aim at equalisation; unless this was achieved there would be an incentive for the opening of new quarries near canals, which could supply inland ports without being liable for duty.

He told Ellis that Lord Penrhyn had gone to London 'on account of this tax', and that Wyatt had given him 'such information as I Thought might be useful'. This information included the following account of vessels carrying slate from Caernarvon and Bangor in 1793:

	Caernarvon	*Bangor*	*Total*
Exports	109 vessels	54 vessels	163
Coastways	140 vessels	226 vessels	366
	249	280	529 [7]

As a result of a meeting of the gentry, clergy, householders and inhabitants of the county of Caernarvon, a memorial was drawn up and presented to William Pitt, the Chancellor of the Exchequer, by Lord Bulkeley, Lord Lieutenant of the county, and Robert Williams, MP for the county.[8] The memorial stated that all vessels cleared coastwise from Caernarvon carried Welsh slate, and that this was also true of most ships cleared coastwise from Bangor. Within the previous ten years quarries had been opened at 'very considerable risk and expense' and on a scale larger than ever before; and the number of vessels had been increased by half. The value of the slate on the bank, after it had been raised and dressed, was under 2s a ton; so to gain a profit the enterprise had to be carried out on a large scale. The industry was currently employing over 2,000 people, and their livelihood was threatened by the tax.

The memorialists complained that the tax gave both inland slate and 'other Materials used in covering Buildings' an 'undue preference at London and other Distant Markets to Welsh Slate carried coastwise to the utter Ruin of a very great Number of Quarrymen, Carriers and Seafaring men and others whose whole Dependence for the Support of themselves and Families is on the Welsh Slate Trade carried Coastwise, there being no Manufacturing or other employment for the lower Class of People in the County of Carnarvon (who are very numerous) but what arises from the Slate Trade.'[9]

After July 1794, despite all protests, a 20 per cent tax was levied on all slate carried coastwise; but as a concession the tax was *ad valorem* and not on tonnage. In practice there were a great many evasions of tax payment; and in an undated note among the Caernarvon Harbour Papers, it was admitted that the duty on slate could not be ascertained. The duty was paid on the full value at the port of delivery, and because of differences in freight it was hard to make an accurate calculation. The note continued: 'No duty is more evaded, not near one half of it is paid so that if a calculation is made and if it should appear to be considerably more than ye Revenue coming from slates . . . into ye Exchequer, it might cause an Enquiry into ye cause of the difference, and ye Evasion found out.'[10] John Evans, writing on 17 May 1825, said that in 1811 'they cheated nearly half the duty'.[11]

An undated memorandum of about 1797 attempted to dissuade the ministers from levying an additional duty on slate sent to Ireland, saying that the 'present Duty' had materially injured the trade 'in consequence of the rigid manner' in which it had been collected. This was probably a deliberate exaggeration; but it is true that Welsh slate producers faced competition from inland quarries which could reach their markets without paying tax. The memorandum went on to say that the number of quarrymen employed had been reduced, and that farmers engaged in carting slates had suffered a loss of employment: as a result, the large number of unemployed had made the poor rates rise steeply. The number of Welsh vessels with slates entering the port of London had been halved 'within the last three years'; and the slaters had been 'obliged to make use of inferior slates from Tavistock and other places, not subject to so much Duty'.[12]

The proposal for a tax on slates sent to Ireland was withdrawn; but in 1805 the existing tax was extended to slates valued at less than 20s per ton, which had hitherto been exempted. In 1809 the *ad valorem* rate was increased from 20 per cent to 35·2 per cent; and although the rate was reduced to 26·4 per cent in 1814 the duty was then levied on tonnage instead of on an *ad valorem* basis.

James Greenfield was the manager of Penrhyn Quarry from 1799 to 1825; and he made a significant contribution to the slate industry by his introduction of galleries or terraces, usually about 55ft deep and 37ft wide, along the side of the mountain. These galleries saved lives by providing a foothold for the quarrymen; and they also saved the slate rock, which no longer fell from a great height to the bottom of the quarry. The galleries were on various levels, and were connected with each other by steps or rope ladders. Tramways were built along the galleries; and as there was a decline from the middle in each direction, wagons could go to either end. This reduced the cost of getting rid of the rubbish; but the installation of galleries was very expensive, and after the depression of 1815 James Greenfield's plan to make additional galleries was not authorised. Instead, men were put to work in higher places at the edge of the quarry. In 1819, during his tour in Wales, Michael Faraday commented with delight on the scene at the quarry 'cliffs', noting: 'Smooth perpendicular planes of slate many many feet in height, depth and width, appeared above and below in all directions, chasms yawned, precipices frowned, and the path which conducted amongst and through these strange places was sometimes on the edge of a slate splinter not many inches wide though raised from the cliffs beneath into mid air.' A system of galleries of about 70ft deep and 36ft wide was adopted in Dinorwic Quarry.[13]

The usual method of making the slates was for a group of workmen acting as partners to take a certain number of square yards of slate rock; from this they made slates at an agreed price, the money being shared among the partners. This arrangement was called a 'bargain'. A bargain usually lasted for a month; but in the early days periods of two months and more were common, and payments were made at intervals of anything from one month to eight months. In 1788 the quarrymen in Penrhyn and Dinorwic Quarries worked in crews, and the wages were calculated from the number of slates according to a graduated series of list prices. In Dinorwic Quarry bargains were set from 21 July 1788 'to continue till the 1st day of February following': three 'Slate Quarrymen' and nine 'good Hands' were to work 'the bottom of Brynglas Quarry' at 13s 6d per thousand for Countesses, 8s per thousand for Ladies and 3s 9d per thousand for Doubles. At the same date another group was given 'liberty to give up this Bargain at the end of first three months if it should not answer to them'.[14]

Cilgwyn were noted for their irregular payments, and on 18 March 1822 the Cilgwyn quarrymen addressed the following petition to the company:

> We the quarry men working in the said quarry, do humbly beseech you will take our humble request into consideration, we all beg your favour to settle the account due Candlemas

last past, and from henceforth we all request it to be paid regularly every six weeks, beginning from Candlemas last past.

We all request to have the Powder for the same price as the other quarries, viz. 8d per pound instead of 1s per pound.

We all from henceforth beg to be paid in Cash, or to have Corn for the market price, when the six weeks would be up.

The slates made by us, are done in the best workmanship and as well as the best in the adjacent Quarries by the Judgement of all the Slate Merchants that has seen them.

The reason of our presenting this our humble request is our poverty, and our Neighbours can't give us Credit, owing to the badness of the times.

The petition was signed by a total of fifty-five men, who were engaged on nine bargains;[15] but it had little effect. On 18 September 1830 the agent Owen Parry wrote to Evan Evans that the Cilgwyn quarrymen were all planning to be at Caernarvon the following Monday. The men had not been paid for twenty-one weeks, and 'all other Quarries pay every half penny, every four weeks in this Neighbourhood'.[16]

The workmen's wage in the first Lord Penrhyn's time was 1s a day, and the overlooker earned 1s 6d a day.[17] About the year 1811 the wages of labourers employed at the port were 2s a day; and by 1814 the quarrymen's wages had risen from 1s 6d to 2s 6d. Wages declined in 1817 to a minimum of 1s 6d, rose in 1825 to 2s 6d, and fell in 1829 to 1s 8d.[18] These wages were slightly above the agricultural wages in the area. At the beginning of the Napoleonic Wars farm labourers in Merionethshire were earning 8d a day in winter and 1s in summer; and in Caernarvonshire the wages were 10d in winter and 1s in summer.[19] Miners on piecework in the copper mines of Amlwch, at about the same date, were earning 'in general from a shilling to twenty pence a day'.[20]

The long intervals between wage settlements meant that 'subsist' or subsistence wages had to be paid to the men to enable them to survive. This money was later deducted from their normal wages. On 12 March 1830 Evan Evans received £30 'to give subsist to the Cilgwyn Quarrymen'; and he divided the sum among fourteen bargains, giving amounts ranging from £5 to 10s.[21] Certain concessions were sometimes given in Cilgwyn Quarry, such as the free use of the company's tools; in November 1820, for example, some men were allowed 'free of expense' to use 'the whimsey, turntree ropes, cibbles and cars'. On 3 June 1822 another group complained 'that they are not allowed the same priviledge as others do in the same Quarry: that they are not allowed anything for opening the Rock before beginning to take out the Slates for which others were Paid.' This was accepted by the men as not being in their contract, but 'the Company hath agreed to set the whimsey to go before they were to begin to work on their Allotment and . . . it was not mentioned in the articles of agreement that they should pay £5 per cent towards Ropes, beside there was no such custom then in the Quarry. However to prevent any ill will we consented to pay 2½ per cent towards the ropes . . . Mr Francis charges on us £6 14s 0d expenses towards setting the whimsey to go before we began to work on a Bargain contrary to our agreement.'[22]

The list prices of slates did not vary very much from year to year, but there were variations in the wages earned from one period to another. This was because a bargain might turn out well or badly, and because piece-rates—meant in theory to facilitate adjustment to the demand for slates—were in practice not very responsive to changes in the price. In addition to the list price, therefore, so much was paid upon every pound's worth of slates produced. This device for equalising the slate bargains was known sometimes as 'price and a half' or 'price and a quarter'. Originally it was a very small amount, beginning at 2s 6d or less in the pound; but it rose to as much as £3. The system was called poundage, and it became built into the industry's wage structure: the poorer the rock, the larger the amount of poundage.[23]

The skilled quarrymen were paid piece-rates; but the rubbishers (ie boys learning to be quarrymen) and labourers were paid by the day. Employers preferred piece-rates which provided an incentive. As Joseph Robinson of Manod Slate Quarry said of his workers in April 1814: 'They will do nothing by the day.'[24]

In the pioneering days of opening up the quarries, discipline was not very strict; and Samuel Holland junior encountered a free and easy atmosphere in Rhiwbryfdir Quarry in the years 1823–4. On 19 February 1823 the men 'stood out for a higher price for making slates . . . but [Holland] soon settled it with them'.

On 7 March he found 'none of the men at work today; except Mr Griffith, Joiner. It was Ffestiniog Fair this day.' On 11 March 'the boatmen returned to their work again and did not succeed in getting the price up.' On 21 March Holland 'told the men they should be fined 2s 6d for every Slate they make undersize.' On 29 March 'the men left off early. Cadwaladar Evans and others went to Parry's Quarry.' On 21 August Holland wrote in his diary: 'Went up to the Quarry. Ffestiniog Fair this day. Had a deal of trouble with some of the men, discharged some; but John Pritchard No 2 and Thomas Pritchard No 3 were very saucy but would not go.' On 7 April 1824 Holland wrote: 'Went up to the Quarry; the men gone a Fox-hunting.' On 19 April he wrote: 'I remained about house all day; Quarrymen were foxhunting.'[25] By the 1830s, however, the organisation of most quarries was improved, and regular hours and rules had been imposed.

Many of the quarry workers had smallholdings, which enabled them to supplement their small wages. These earnings were subject to deductions, usually at 8d per lb, for gunpowder: this was first used in Penrhyn Quarry in 1780. The men also had to pay for fuse; and in the mines they paid for candles. Lord Penrhyn founded a Benefit Club at Penrhyn Quarry in 1787; and this was re-established in 1825. The men's subscriptions were 7d per month, and the sickness or accident benefit was 3s 6d per week. The doctor was appointed by Lord Penrhyn; and the club was run by the agent and one of the managers. Membership was said to be voluntary; but monthly subscriptions were none the less deducted from wages at the quarry office.[26]

The isolated situations of most of the quarries, and the almost total lack of public transport, made the supply of provisions a difficult matter. Many of the quarry owners supplied corn to their workers; and this did not necessarily mean that the men were exploited through high prices. The manager of Penrhyn Quarry organised bulk buying of barley in 1812; in July of that year a total of £1,876 5s 4d was paid for barley, including 'cartage and portage'. The loss sustained on the sale of this barley was £221 15s 10d.[27] At the Dinorwic Quarries in 1818, a loss of £13 5s 3½d 'by Corn purchased for the Quarrymen' was noted in the accounts.[28] At the Cilgwyn Quarries the truck system operated in its worst form. In September 1823 the company paid £13 1s 0d for 'Corn to Quarrymen'; and the money was then deducted from the men's wages. Potatoes, wheat and coal were also supplied by the company, which issued tickets to the quarry workers instead of cash so that they could buy these goods.[29] The Welsh Slate Mining Company opened a shop for their workers at Ffestiniog in September 1826. This was described as:

> A very large retail flour, butter and bacon warehouse at Rhiwbryfdir Quarry, Ffestiniog (under the management of Mr Homfray) for the accommodation of the Quarrymen, which will save them the trouble and expence of going to market elsewhere for those articles.

The *North Wales Gazette* asked: 'Would not a bakehouse and small beer brewery have been an additional accommodation to them?'[30] The truck system lent itself to abuse; and there were complaints from the men that they were having to pay excessive prices for goods. In the slate industry, however, the system was not widely operated in its most objectionable form.

The profits of most of the smaller slate concerns, such as the Cilgwyn Quarries, were comparatively small. Many were often in serious financial difficulties, and were unable to pay their workers. Penybryn Slate Quarry, in the parish of Llandwrog in Caernarvonshire, was worked from 1808 by a partnership consisting of William Turner, Richard Garnons, Hugh Jones and William Wynne. Richard Garnons had leased the quarry for thirty years, with one-eighth of the value of the slates to be paid to him.[31] By February 1818 the concern was in debt to the sum of £920 1s 1d. The only profits during the previous decade had been in 1812, when £400 was 'shared between the Partners', and in 1813, when William Wynne and Richard Garnons had £100 each.[32] Hafodlas Quarry at Llanllyfni in Caernarvonshire, made a profit of £167 0s 6½d in the year 1809;[33] and Lord Newborough's slate quarries at Glynrhonwy on the western side of Llyn Padarn realised a profit of £986 3s 9½d in the year 1827.[34] Between 30 May 1823 and 8 February 1825, however, Llwydgoed Bach Quarry in Llanllyfni made a loss of £319 3s 5½d.[35]

The number of men employed at the Glynrhonwy Quarries in 1826 was 197.[36]

The number employed in some of the Nantlle Quarries was very much smaller, although there were often fluctuations in the size of these concerns. On 15 April 1823 John Evans was informed that no quarrymen at all were working at Manod Slate Quarry.[37]

Only 8 workmen were employed in 1824 at Moeltryfan Quarry, which had run into serious trouble with flooding.[38] In 1821 the number employed at Penrhyn Quarry was between 700 and 800 and by 1826 this had increased to 1,200.[39] The Dinorwic Quarries employed a smaller number at that date; but by 1832 they were employing 800 men.[40]

The tax on slate sent coastwise gave an incentive to quarry owners to find overseas markets. There was little export of slates to the continent before the 1840s, but trade with America had begun by the end of the eighteenth century. Arthur Aikin on his visit to Caernarvon observed 'an American ship of about 400 tons, that had landed tar, potash and other articles, and was taking in a cargo of slates, flannels and a fine sort of ochre . . . found in Anglesey.'[41] In 1804 a slate merchant in Liverpool wrote to John Evans:

> Am weekly conversing with American Agents about orders for Welsh Ladies . . . yet not obtained a demand for such hardware. Doubles are seldom wanted being too small

size, or should be glad to hand you £300 for your stock, so as to clear the way with Cilgwyn Quarries.[42]

In 1816 John Evans was still investigating the possibilities of the export trade; and on 16 July he received a somewhat discouraging letter from two Liverpool merchants, Lodges and Tooth. They told him that there was no general demand for slates in America, but continued:

> The export of them has been chiefly to Boston where a Slate Law has been lately made to Slate all buildings—and in consequence we have had orders to ship three Cargoes and other people have been also shipping to the same port—few howsoever are ship'd here and you would be enabled to get the best information as to the extent of the export of Slates from the Customs houses of your own place at Bangor or Beaumaris. Its probable a Cargo of Slates would still do well at Boston . . . we know nothing about the Newfoundland Trade . . . There is no export of Slates from *hence* to the Northern European ports—we know not to what extent they are used there or whether at all—but we don't think any good would be done by a Casual Cargo sent to any of the places we have in view.[43]

By March 1817 the *North Wales Gazette* was exultantly congratulating 'the country upon the prospect of a very great increase to its commercial interests, in what may be considered the staple articles in these counties—There are now three large vessels taking in cargoes of SLATES for North America, and several more are expected. Two are loading at the port of Velin Heli, from the quarries of T. A. Smith Esq and the third from Port Penrhyn.'[44]

In the following month the same newspaper announced that 3 large ships were ready to sail for America, 'laden with slates from the Penrhyn Quarries'.[45]

The early export trade was a risky matter. Fluctuations in

demand made it difficult for slate merchants to procure detailed information about the state of the market; and the difficulties were increased by frequent delays in the transporting of slate, and by the need to give long credit to buyers. The American trade was therefore regarded as speculative, and merchants preferred the safer home market. A document signed on 30 October 1816 by William Hughes, a farmer in the parish of Llanwnda, and Robert Hughes, an insurance agent, illustrates the protective measures that could be taken. The two men declared that 'in the event of the Hafodlas Slate Company loading the brig *George*, Captain Lewis, with slates for America on Account', they would back the transaction: William Hughes bound himself for the sum of £500 'to execute a regular Bond for the payment of the Amount of the Invoice of the said Cargo'; and Robert Hughes undertook to 'insure the value of the said cargo and to deliver up the policy of insurance to Mr George Bettis as a further security previous to the sailing of the aforesaid vessel.'[46] Insurance became increasingly important with slate cargoes even for the home trade. In March 1824 Joseph Williams, a shipbroker in London, wrote to John Evans asking him to get a policy of insurance on the sloop *Cilgwyn* for a voyage from Caernarvon via the Isle of Wight to London. The vessel was to be insured for £350 and the cargo for £155.[47]

Trading conditions in the early nineteenth century were often very unsatisfactory. Slate merchants were of two classes, those who owned their own vessels and those who chartered vessels; and both types took the cargoes on three or four months credit, usually trying to sell them at a profit to slate brokers in various seaports of England and Wales.[48] The slate concerns which owned their own vessels included those of Penrhyn, Dinorwic, Cilgwyn and Ffestiniog. Some of the trading difficulties arose because of attempts to cheat the customs officers of slate duty. This was the cause of trouble in December 1820, when the shipping agent of William Turner's Ffestiniog quarry consigned a cargo of slates to London to a merchant called E. O. Jones. The captain of the brig *New Liberty* discharged part of the cargo in the absence of the customs officers, and deliberately miscounted it. This was with the connivance of the merchant, who had been 'for years cheating the Crown of the Duty for slates in the wholesale way'; but on this occasion a former employee of the captain, who bore him a grudge, laid information against him at the customs office. John Evans advised Jones to be 'careful in future', and to make his affidavit stating such truth as it was 'prudent' to tell. 'All the truth', he continued, 'would not answer your purpose.'[49]

Other difficulties arose when a customer rejected a consignment of slates and refused to pay for it. In 1823 Samuel Anderson, a Glasgow customer of William Turner's, refused to accept a cargo which had been selected and approved by the captain who was his agent. Anderson complained that the slates were not three-eights of an inch thick as required for the roofing of Carstairs Castle, and said that he had been 'ordered to take and cart my slates back to Glasgow or to do with them what I please'. William Turner replied that the slates sent were the thickest they made. An action was taken against Anderson for non-payment of the slates, but Anderson settled the matter out of court by paying the debt.[50]

Sometimes merchants refused to pay for slates they had received. William Turner employed his brother-in-law George Ford of London as his slate merchant from 1808 till 1817. When Ford died, he owed Turner £1,449; and this resulted in a chancery action between the Turners and the Fords, which lasted until 1828.[51] Respectable slate merchants resented the speculating merchants who, according to a letter written in 1824 by Thomas Hoskins of Gosport, 'ran about the country underselling regular merchants yet were supplied on more favourable terms by the quarry owners'. Hoskins was determined 'not to countenance these Men as they do great injury to the Trade altogether, and the Merchants along this Coast mean to withdraw their Dealings from any Quarry who send out these runners about this Country.' This hawking of slates was deprecated by Hoskins, especially as some of his established customers were being affected. He offered his services to John Evans at six months' credit, giving references to respectable timber houses in

London, and suggesting that Evans's sloop *Cilgwyn* might be sent back loaded with foreign timber.[52]

Shipping expanded as the slate industry developed. There were only a few private companies, and no single person or joint stock company was prominent in the industry. Samuel Samuels, who died in 1821, was an important shipwright in Caernarvon and was one of John Evans's clients.[53] Most of the ships used for the transport of slate, however, were owned by farmers, quarrymen, sea captains, shopkeepers and professional men who lived near the shipbuilding ports. These ports included Aberdovey, Barmouth, Conway, Portmadoc, Port Dinorwic, Pwllheli, Nevin, Caernarvon and Bangor. Most of the slate vessels were built of wood, and the capital for their construction was provided by a large number of shareholders who invested small sums and received small dividends in return.[54] Shares in these vessels were frequently sold by auction. The following advertisement is typical of these sales:

> To be sold by Auction at the Crown Tavern, in the town of Carnarvon, on Friday the 12th Day of September, 1823, between the hours of 3 and 5 o'clock in the afternoon . . .
>
> Six Sixteenths Parts or Shares
> of that New Strong Built,
> And Fast Sailing Smack,
> called the
> Carnarvon Packet,
> of the Port of Carnarvon,
> Burthen per Register 53 2/94 Tons.
> Griffith Jones, Late Master.
>
> This Vessel now lies in the Port of Carnarvon, has been out but one Season, and is well calculated for the Coasting or any other Trade to which her burthen is suited.[55]

John Evans, besides having a share in the sloop *Cilgwyn*, had interests in other vessels: one of these was a brig of 240 tons, 'a fast Sailing Vessel' which he hoped to send 'to an American port'.[56]

Page 103 (*above*) Llechwedd Quarry, showing a slate dressing mill with a line of saw tables; (*below*) Llechwedd Quarry in 1895, showing two slate mills, three waterwheels, and a steam locomotive and trucks of the Festiniog Railway

Page 104 (*above*) Dinorwic Quarry workshop, with a view of the circular slate dressing machines; (*below*) the Dorothea Engine House in 1907, containing the Cornish beam pumping engine

Payment for slates in 1824 was usually by bill of exchange at 3 months from the date of the invoice, or by bank notes often with a discount of 2½ per cent.[57] The opening of the copper mines in Anglesey and the slate quarries in Caernarvonshire and Merionethshire led to the establishment of banks in North Wales. In 1792 the Chester and North Wales Bank was set up, with capital largely from Anglesey copper mines; and in 1812 and 1822 it opened branches in Caernarvon and Bangor. The bank of Thomas and Hugh Jones was established at Dolgellau in 1803; and Hugh Jones became the partner of Turner and Casson in their Ffestiniog slate quarry.[58]

The use of gunpowder added considerably to the dangers of slate-quarrying, especially in the early nineteenth century when rules for its use had not been formulated. A 'shocking accident' at Penrhyn Quarry in July 1810 was reported as follows:

> In blasting the slates, an instrument called a stamper, which is 30in long and 2½in in circumference, is used to ram down the charge of gunpowder; it is supposed the friction of working out the stamper, produced a spark, which communicated and caused a sudden explosion, driving this thick iron rod up the muscles of the workman's arm, entered through to the neck, advancing nearly 8in beyond. His death was instantaneous.[59]

In 1821 the Rev P. B. Williams regretted that the benefits to the 'labouring poor' of employment provided by the slate industry should be diminished by accidents, such as 'the sudden explosions of Charges of Gunpowder, the falling of stones, rubbish and fragments of Rocks etc, and breaking of ropes, whereby many of the workmen are lamed and maimed, and others lose their sight, and thus become chargeable to different Parishes.' A description of the work at Assheton Smith's early quarries of Allt Ddu and Clogwen y Gigfran indicates the dangers. Many of the men descended fifteen or twenty yards 'by the assistance of two Ropes (one about their middle, and the other in their hands) to a small ledge, over a dreadful precipice, where they continue engaged for many hours, in boring, or detaching considerable fragments from the main rock, and ascend again in the same manner.'[60]

Mechanisation of the slate industry made some progress in the early capitalist period. The quarrymen, however, still had to detach the slate block from the quarry face, using a process known as 'pillaring' to exploit the slate's tendency to break vertically at right angles to the direction of cleavage; and after that they had to cleave and divide the blocks, to split them and to dress the slates. The first operation, the quarrying of the blocks, was achieved by drilling or boring a hole along the cleavage of the rock and inserting a charge of gunpowder. Usually two of the partners in a bargain were quarrymen responsible for obtaining the blocks, and the remainder were splitters and dressers. As many slates as possible were obtained from the block by careful splitting, pillaring and the avoidance of natural blemishes. The small blocks were then split into slates, the tools used being thin chisels or 'splitters' and wooden mallets bound with iron. The irregularly-shaped slates were then taken to the dresser, who placed the slate block on a bench or 'carrier' and squared it with a long knife.[61]

New methods of production were introduced in some quarries during the 1820s. Horse whimseys, which were used to hoist the slates, were in operation in that decade at Cilgwyn; and turntrees or winches were being used at the same quarry in 1827. Another innovation was the water balance. An iron tank was fixed upon wheels and put upon an incline, so that when filled with water it was able by its greater weight to raise slate blocks; when the tank reached the bottom, it could be emptied. A water balance was installed in 1829 at Tallysarn Quarry in Nantlle; it was used to lift slates in a vertical shaft.[62]

The first steam pump in the history of the North Wales industry had been erected about 1807 at Hafodlas Quarry. It cost £1,000 'or thereabouts'; and it was advertised for sale in 1811, when the quarry was up for auction. The machine was used to keep the quarry free from water; but in August 1817, 'after and in consequence of a heavy and incessant fall of rain for several days a certain Bank adjoining the said quarry, and on which the said Steam Engine and two whimseys were erected, gave way and precipitated the whole together with the said Steam Engine and whimseys to the bottom of the said quarry and to the depth of sixty or seventy yards and thereby most of the cast iron pumps belonging to the said Engine and whimseys were shattered.'[63] No one was injured; but it was alleged that the accident had been caused by the negligence of the quarry manager, George Bettiss. It was proved, at any rate, that Bettiss had kept very inaccurate accounts; more than £2,000, which had been received by him for slates, was not accounted for in the quarry books for 1812–17. After 1817 the quarry was closed for some years; and in July 1828 it was again put up for auction.[64]

The cost of transporting the slates was often a key factor in the success or failure of a slate concern. Shipping rates fluctuated, increasing by about a third between 1790 and 1820; but these rates were less important than the cost of carriage from the quarries to the ports. The extant slate carriers' accounts for Cilgwyn Quarry give the names of carriers who drove carts to Caernarvon: most of them seem to have been farmers and cottagers. The carriage prices paid by the quarry owners in 1822–3 were as follows:

Tons Queens and Rags	8s	per Ton.
Duchesses	20s	per mille.
1st countesses	14s	,, ,,
2nd countesses	11s	,, ,,
1st Ladies	8s	,, ,,
2nd Ladies	6s 3d	,, ,,
1st Doubles	4s	,, ,,
Singles	2s 6d	,, ,, [65]

In 1823 John Evans, on behalf of the Cilgwyn Company, gave an undertaking to the toll-keeper at Pont Seiont that he would pay the toll for nine slate carriers every two months.[66] In the following year Mary Ellis, one of the slate carriers, was warned that if she did not pay 7s 0d which she owed for tolls at Hendre Turnpike Gate, plus 3s 6d costs, she would be prosecuted.[67] The carriers were usually engaged for the year; and in 1802, when a list was made out, ninety-one carriers were working for the Cilgwyn Company. Most of these owned one cart each, and went to Caernarvon; but a few went to Voryd, and there were four carriers who owned two carts each. The average load was 18cwt; and the usual rate was 10s or 11s per ton.[68] The slate carts did not improve the road surfaces; and in May 1824 it was reported that Evan Evans had retained a total of £10 6s 10d from sixteen carriers to pay for road repairs.[69]

Another problem was the fact that the carriers often needed their carts for farmwork, and in any case they were slow and expensive. Iron tramways were cheaper and quicker, and quarry owners were very anxious to promote them wherever possible. There were often long delays, however, because of opposition from the carriers whose livelihood was endangered.

In February 1813 the proprietors of Cilgwyn, Penybryn, and Talysarn Quarries met in Caernarvon 'to take into Consideration the Utility of applying for an Act of Parliament to authorize them to make an Iron Tram or Railway from their respective quarries to the port of Caernarvon . . . for the purpose of Carrying their Slates to Market to avoid the present exorbitant prices paid for Carriage by Carts.' The quarry proprietors had already employed an engineer to make a survey and estimates for the proposed line.[70] It was established that there would be a saving of 10s per ton in transport costs: the cost by cart was put at 14s and the cost by rail at 4s, including interest and repairs. It was also hoped that the railway would reduce the allowance necessary for breakages, which was then 1cwt in every 20. By 1814 Hafodlas Quarry was included in the proposal; and it was suggested that the four quarries should contribute four equal shares of £3,000 each.[71] The railway was finally authorised on 20 May 1825; and it was built by a joint stock company. The resident engineer was Robert Williams; and the shareholders included Richard Garnons, J. W. Trevor, W. Potter and W. Wynne Williams. It was a horse tramway with a gauge of 3½ft; and it ran from Gloddfarlon Quarry in the parish of Llandwrog to Caernarvon, a distance of about seven miles. It was opened in 1828.[72]

Another horse tramway was built by Thomas Assheton Smith across Fachwen to Port Dinorwic, a distance of about six miles: this was completed in 1824 at a cost of over £9,000,[73] and did much to speed up output and reduce transport costs. It went across a ridge, however, and had a gauge of 1ft 11½in. As the slate business expanded it was found increasingly inadequate; and a new railway was accordingly made in 1841–2 along the shore of Llyn Padarn.[74]

Development of the slate ports had begun in the 1790s with the transformation of Abercegin into Port Penrhyn; the short distance of six miles to this port from Penrhyn Quarry was a great asset, and added to the quarry's advantages. The port contained a sawing mill where the slate was sawn in to gravestones and 'chimney pieces', and a large factory, where school or writing slates were polished and framed. These writing slates were not subject to the slate duty; and there was a temptation, sometimes not resisted, to claim that slates for roofing and other purposes were school slates. In a letter to William Turner on 20 October 1812, George Ford wrote: 'Please to concine all the flags to me and call them writing slates and then they will be free of Duty.'[75] In 1813 907 boxes of writing slates were shipped from Port Penrhyn. The number fell to 699 in 1815, but rose to 1,128 in 1816; in 1819 it fell again to 802.[76]

Port Dinorwic, known at first as Moel y Don or Velin heli, was improved and enlarged in the early 1790s and again in 1828. In March 1793 John Byrne wrote to Hugh Ellis: 'We go on briskly with Moel y don Quay, which I hope will soon be finished.'[77] A road from the slate quarries to Port Dinorwic was begun in 1809, and notices issued in 1818–19 insisted that this road was private: 'If you travel with your Horses, Carts or otherwise upon and use the Road lately made by the said T. A. Smith, Thomas Wright, William Turner and Hugh Jones leading from the Dinorwic Slate Quarries to the Shipping Place at Moel y don, through the private property of him the said T. A. Smith situate lying and being in the Parish of Llanddeiniolen in the County of Caernarvon you will be proceeded against as a wilful Trespasser.'[78]

In the eighteenth century slates from Nantlle quarries had

sometimes been sent to Voryd; but in the early nineteenth century they were taken mainly by cart along the turnpike road to Caernarvon. The slate quay at Caernarvon was built about 1803. By 1821 it had been 'extended, and rendered more spacious and convenient'. It was said that 'scores, if not hundreds of Waggons and Carts' were engaged 'in bringing down the production of the Quarries to this town'. The slates were exported 'to various parts of the world'; and the average annual value of exports was about £50,000.[79] The construction of the Nantlle Railway enabled the quarries to maintain a constant supply of slates to the port; and this stimulated the export trade.

The development of Portmadoc, the port for the Ffestiniog quarries, was largely the work of William Alexander Madocks. Madocks, who was born in 1773, inherited a fortune from his father; and in 1798 he purchased the estate of Tanyrallt near Penmorfa Marsh. The estuary of the Glaslyn divided Caernarvonshire and Merionethshire; and the sands of the Traeth Mawr were notorious for their treacherous nature at low tide. It was Madocks who, following an idea proposed by Sir John Wynn of Gwydir in 1625, succeeded in obtaining an Act of Parliament in 1807 which vested in him 'the tract of sands situate on the Estuary called Traeth Mawr . . . extending from Pont Aberglaslyn to the Point of Gest'. He built an embankment to protect the sand from the sea; and by 1810 he had also built a one-mile railway along this embankment.

The engineer was Thomas Payne of Warwickshire, and about 5,000 acres were reclaimed from the sea. The work was completed, amid much local rejoicing, in 1811. A gale in February 1812 made a disastrous breach in the embankment, but there was widespread response to appeals for help. A subscription fund was set up in 1812; and Percy Bysshe Shelley, who was then living in the neighbourhood, promised £100 and spoke at a fund-raising meeting in Beaumaris.[80] The publicity given by Shelley to the cause was undoubtedly helpful, although early in 1813 he abandoned the project and left Tremadoc. Thomas Love Peacock, on the other hand, lamented in *Headlong Hall* that a spectacular bay had been converted into a 'sandy waste'.[81]

At this stage the port was still Ynys Cyngar, a mile west of the harbour of Portmadoc; and the quarries continued to send down slates by carts and wagons to the Traeth Bach. A number of quays existed on the River Dwyryd, such as Cemlyn, Tyddyn Isaf and Cei Newydd, and as the carts arrived at the quayside, men, known locally as 'Philistines', loaded the slates into small boats of about six tons burden. At Ynys Cyngar the Philistines transferred the cargo to the sailing ships, while they returned to the quays on the Dwyryd.[82] Madocks realised that these facilities were inadequate for the expanding quarries north of Ffestiniog; and he obtained in 1821 an Act of Parliament which authorised him to build a harbour, to be called Portmadoc, at the west end of the embankment. The harbour was completed in 1824; and in 1825 about 12,000 tons of slate were exported from it. The tramway which ran along the top of the embankment was incorporated later in the Festiniog Railway. The town of Tremadoc was a model town built by Madocks on land reclaimed from the sea, and called after him. Madocks himself died in comparative poverty in Paris in September 1828, having 'spent a princely fortune in useful improvements'.[83]

Conway was the port for the slate region comprising Dolwyddelan, Capel Curig, Bettws y Coed and the Conway Valley; but its importance was not significantly increased by these quarries, most of which were short-lived ventures. In October 1814, for example, Dolwyddelan Quarry was let for 3 years to 3 quarrymen. The conditions of letting included the right to part of Fany Castel 'for the purpose of accommodating themselves and the married workmen with grass and hay for milch cows and gardens'. For this privilege the 3 quarrymen were to pay 'only 1s per annum as a nominal rent'; and they were to employ at least 20 workmen. The agreement was not, however, maintained for the proposed 3 years: 2 of the 3 partners withdrew after only 1 month, and in 1816 the third brought a successful law-suit for settlement of what he was owed.[84] The quarries of Tan Rhiw and Blaen y Cwm were advertised for sale in July 1826; and Tan Rhiw was up for sale again in 1828.[85] The inferior quality of the slate rock, plus the heavy cost of transporting the slates to Trefriw and then to Conway, were the main factors operating against the success of quarries near Dolwyddelan and Penmachno.

The selling price of most slate rose by about 20 per cent during the early years of the nineteenth century, as the following figures for Dinorwic Quarry show:

	1808	*1825*	*1826*
Duchesses per Thousand	£5 5s od	£7 os od	£7 15s od
Countesses ,, ,,	£3 3s od	£4 10s od	£5 5s od
Ladies ,, ,,	£1 18s od	£2 os od	£2 5s od
Doubles ,, ,,	18s od	16s od	18s od
Queens per Ton	£2 2s od	£2 5s od	£2 10s od
Rags ,, ,,	£1 16s od	£1 12s od	£1 15s od

The prices for best slates in 1825 were roughly double those of 1798, and the prices in 1826 were said to be the same as at Penrhyn and the Ffestiniog Quarries. Prices at Talysarn Quarry in 1819 were identical with those at Dinorwic in 1825. Prices in 1826 at Glynrhonwy Quarry, however, were considerably lower.[86]

Duchesses per Thousand	£5 17s od
Countesses ,, ,,	£3 17s 3d
Ladies ,, ,,	£1 12s 7½d
Doubles ,, ,,	16s od
Rags per Ton	£1 8s od

The 1820s were a period of speculation in the quarries, of litigation between slate merchant and quarry owners and between landlords and lessees, and of strife between quarry owners and quarrymen. Relations at Penrhyn Quarry had become very unsatisfactory by 1825, and on 19 January 1825 James Greenfield was found drowned. In March 1826 discontent erupted; and the quarry came to a halt 'in consequence of about 150 men from the lower Quarry striking from their work, who proceeded through the Quarry and drove the rest off.' Dawkins Pennant refused to meet the men; and on 22 March he issued a statement. Those who did not wish to return to work were told to go to the quarry in order that they might be paid off; and those who wished to return to work 'on the same terms as heretofore' were requested 'immediately to attend their several employments'. It was promised that when the men had returned to work Benjamin Wyatt, the new agent, would inquire into their complaints.[87] On 24 March the quarrymen addressed a petition in Welsh to Dawkins Pennant, asking for higher wages and fairer treatment. They complained that wages varied from 17s a month to £5–£6 a month; and they asked for a minimum wage of 3s a day. They also criticised one of the overseers.[88] The strike achieved nothing; and negotiations were broken off when a reference to James Greenfield's death was interpreted as implying a threat to the managers of the quarries.

In Dinorwic Quarry, at about the same date, a new rule was introduced whereby the men were to continue working on Saturday until 4 pm instead of 1 pm. The managers stationed themselves at the main exits, but all the men started to leave together at 1 pm, and walked past the supervisors with 'good-humoured contempt'. No further effort was made to alter established customs, and for almost fifty years thereafter amicable relations existed between the men and their employers.[89]

The quarrymen's working conditions were very bad in this period. Their work was exhausting and dangerous; and it was often performed in bleak, exposed situations on mountain slopes. In some quarries there were no sheds in which the men could split and dress their slates during wet weather; and few of the early capitalists gave much thought to their workers' welfare. On 15 October 1804 William Pritchard, who had been sent by John Evans to survey Manod Quarry, wrote to him suggesting a more humane approach to the workmen: 'In my simple opinion as you intend to carry the work on, you ought to have erected some kind of huts against the winter season that the workmen might have a comfortable place to eat the bit they bring with 'em, as it's impossible for them to stand it out in cold weather; as it is, we had hail and snow there last Thursday and Friday.'[90] This quarry, on the southern slopes of Manod Mawr, was at an altitude of 1,600ft.

Many of the quarrymen walked long distances to work. Some

came to Ffestiniog over the mountains from Beddgelert; and in wet weather, when they arrived soaked, no drying facilities were available. It became common practice for some of the men to stay in lodging-houses or 'barracks' during the week and return home only at weekends. The barracks were described at Dinorwic as 'poor little houses'; and in 1873 they were replaced by new ones.[91] Houses for the quarrymen were built, however, by Lord Penrhyn and Thomas Assheton Smith. Lord Penrhyn created the model village of Llandegai with its neat houses, and he also built a number of 'neat slated cottages for the accommodation of the labourers in the valley and along the sides of the River Ogwen'.[92] Thomas Assheton Smith had granted leases on the Vaynol estate to almost all his tenants for twenty-one years; but when the leases expired in 1799 and the tenants had not significantly improved their holdings, he determined not to repeat the experiment.[93] He built a number of houses when the railway was being constructed; and some cottages were also erected near Dolbadarn Castle.[94] On 27 February 1819 Thomas Wright wrote to William Turner, revealing his own feelings towards the men: 'I shall be glad to hear that you have made the Quarrymen easy by raising their Wages, and you must be liberal in your Charity to the Families of men killed and hurt in the Quarries.'[95]

Thomas Assheton Smith died at Tedworth on 12 May 1828. He was described as 'a kind, and an honourable and upright man'.[96] His son, Thomas Assheton Smith (1776–1858) assumed complete control of the quarry; but William Turner continued to act as manager. The new owner was to be remembered rather as a foxhunter than as an industrial entrepreneur; but—perhaps because of this love of sport shared by the quarry workers—he managed to retain the goodwill of his quarrymen for most of his life.

On 15 May 1828 he wrote to the Vaynol estate-agent John Millington, indicating the problems he was taking over:

> I am glad to receive yr. Letter this morning in which you mention improvement in ye Demand for Slates. It is very necessary there shd. be a good receipt, as altho' my Father

> has done *most honourably*, yet he has left all ready money in Bank to my Sisters. He has left all arrears to me and all money coming . . . I don't even know the Rent days . . . Let me know all this . . . Tell me too in yr. next Letter . . . the names of . . . London Slate merchants with whom you have *ever* had Dealings. I wrote a long and very strong Letter to ye Treasury a week since, and as yet have had no answer, stating that our Quarries must stop, men thrown out of Employ—and coasting Vessels in numbers laid up—if they don't in some way remit the Duties . . . I will have a personal Interview with the Chancellor of ye Exchequer, with whom I am intimate when I go to Town.[97]

Millington wrote to Assheton Smith that he had visited a number of slate merchants, including Evan Roberts 'one of the most extensive dealers in that Trade in London'; Roberts and others had promised that they would be customers of Dinorwic Quarry.[98] Having taken over the business side of the quarry, Thomas Assheton Smith was urged by Peter Bayley Williams, the rector of Llanrug and Llanberis, to look after the welfare of his workers. On 29 June 1829 the rector wrote telling him about a Friendly Society or Savings Club which had recently been established at Llanberis, and asking if Thomas Assheton Smith would give or rent them a piece of land near Dolbadarn Castle so that they could build a meeting-place there. The rector strongly recommended the society: it would provide benefits to cover illnesses and accidents, which were 'of very frequent occurrence at the Slate Quarries', and it would also encourage the men to be independent and not rely on parish relief. The rector also informed the new quarry owner that his workers were complaining about the quarry shops, as prices there were above those in Caernarvon. The situation was a complicated one, as the shopkeepers denied the allegations and claimed that the men were combining to injure their reputations. The rector was well aware that it was 'a very delicate object to interfere between a gentleman and his workmen'; but he described the quarrymen as an 'honest sober industrious and contented set of men (taking them as a body in the aggregate)' and declared that they 'disliked drunkenness and immorality—and excess and irregularity of every description'. He suggested that a few shops might be erected near the Dolbadarn Inn: this would save the inhabitants of Llanberis from having to 'climb up the side of a steep mountain a full mile to the Quarry shop'.[99]

This picture of the sober Llanberis quarrymen is impressive; but it had been suggested in March 1828 that two young men might be sworn in as constables 'as some young Quarrymen are frequently very riotous at the Inn'.[100] In July 1832 John Millington told Thomas Assheton Smith that 'Holy Thursday' was an annual holiday, when the men went to church and elected 'new officers for their Benefit Club'.[101]

The slate industry was in a depressed state when Thomas Assheton Smith inherited his quarries, and he lost no time in using his influence to get the slate duty abolished. Lord Newborough sent a circular to all quarry owners in December 1830, summoning them to a meeting at the Uxbridge Arms Hotel in Caernarvon 'for the purpose of taking into consideration the unprecedented depression of the Welsh Slate Trade, and the propriety of petitioning Parliament to have Wales put upon an equality with England and Ireland as regards the duty affecting it.'[102] This was only one of many meetings; and as a result of numerous petitions, including those got up by Samuel Holland, the slate duty and that on coal were repealed in 1831.[103]

The early capitalist period was marked by a great variety of undertakings, ranging from the large-scale concerns at Penrhyn and Dinorwic to primitive, poorly-financed organisations such as those on the Commons of Manod and Cefn Du. By 1830, however, the slate quarrymen had emerged as a distinctive group. The population of Caernarvonshire rose from 41,521 in 1801 to 66,448 in 1831; and that of Merionethshire rose from 29,506 in 1801 to 35,315 in 1831.

6

EXPANSION IN OUTPUT, 1831–1882

The population of England and Wales increased from 8·8 million in 1801 to 25·9 million in 1881. The demand for roofing slates fluctuated as the building trade prospered and declined; but on average it rose, so that the supply was outstripped and delays in receiving orders were regarded as inevitable. From 1831 to about 1882, the slate industry continued to expand; and these years of expansion were later seen as the industry's Golden Age.

For two years following the repeal of the slate tax, the duty on tiles was retained; and this gave an advantage to the slate producers. Even during the Napoleonic Wars, however, the high price of timber had encouraged builders to use the lighter slates, which needed less roofing timber than the tiles and stone slates. Thanks to the growth of new centres of industry and the expansion of established ones, the building trade in the early 1830s had difficulty in keeping pace with the demand for new factories and housing.[1]

In 1833 the tile duty was abolished, but the building boom had by then begun; it was checked only by the monetary troubles of the years 1836–9. In 1832, however, it was not tiles which were disturbing Thomas Assheton Smith and other quarry owners, but the use of zinc as a roofing material. On 26 June Assheton Smith wrote to John Millington on this subject:

> You will have heard of the alteration Government are making with the Duties on Import. I fear the Slate Trade will be materially injured should marble, zinc, iron, copper and other articles be allowed to come at the lower duties pro-

> posed. Zinc is coming into use for roofing. In Russia and America it is very much used and approved of and has also been lately used by the Corporation of Liverpool in preference to slates for covering sheds adjoining their Docks. The duty *now* upon it is £10 per ton and it is proposed to reduce it to £2. I have not yet been able to ascertain the value of it but have been informed that even at £10 p. Ton it will in many places be as cheap as slates for roofing.[2]

Proposals for the reduction of these import duties had been introduced by Huskisson, who was fatally injured at the opening of the Liverpool & Manchester Railway in 1830. When they came into force, they had a marked effect on the Welsh copper industry; and that effect became still greater after 1848, when the duties were abolished completely by Lord John Russell. Competition from zinc and iron for roofing was slight, however, until the end of the nineteenth century. At that time, corrugated iron sheeting began to replace slate, thatch and tiles on farm-buildings: the low cost of this material blinded landlords to its poor insulating properties, and to its need for protection against rust.[3]

By 1835 the British canal system was virtually complete.[4] Cheaper transport had reduced the costs of the slate industry, especially after the end of the Napoleonic Wars when competition from improved road transport persuaded some of the canal companies to reduce their rates. In 1817, for example, the Glasgow, Paisley & Johnstone Canal had reduced its charges for goods including slates passing between Glasgow and Johnstone, fixing them at 5d or 5½d per cwt.[5] By the 1830s, however, the canals were faced with competition from the railways, and the decline of their prosperity began. Freight charges on the canals were reduced; and the development of the railways brought slates cheaply to inland towns. Shrewsbury, for example, used slates in preference to tiles after the arrival of the railway; and regional traditions in building began to disappear as bricks and Welsh slates were carried all over the country.[6] Apart from the Chester & Holyhead Railway and a few local lines, however, Wales itself was still without railway communication in 1844; and it was not until the Cambrian and Central Wales lines reached the counties of Merioneth, Montgomery and Radnor in the 1860s that these remote districts were brought into touch with other parts.[7]

The short local lines were of great importance to the slate industry. The most successful was the Festiniog Railway, which is today an important tourist attraction. After the failure of Madocks' early attempt to connect the Ffestiniog slate quarries and the harbour by rail, Samuel Holland and his father managed to get the scheme under way. The first bill was strongly opposed by farmers and boatmen, and was thrown out at the committee stage; but in May 1832 the railway was authorised by Parliament. Henry Archer was employed as manager, and James Spooner and Thomas Pritchard were called in as engineers. The first stone was laid by William Oakeley near Creua on 26 February 1833. In June 1833 Samuel Holland junior and Henry Archer, the director of the Festiniog Railway Company, concluded an agreement whereby the company undertook to make the inclines necessary to enable slates to be brought down from Holland's Quarry to the railway. Holland was to pay the cost of this over a period of two years; and he also engaged to send all his slates by the railway to Portmadoc. As Holland was the first quarry proprietor to make arrangements with them for the carriage of slates, the company agreed to give him £250 towards making the inclines. The railway company also agreed to accept 6s per ton for the carriage of slates.[8] The gauge of the railway was 1ft 11½in, and it was about 14 miles in length.

The railway was completed on 20 April 1836, when there was a 'grand opening day'. A train carrying slate and people, including the railway workmen and quarrymen, travelled over the line. Samuel Holland was one of the passengers, and he subsequently described the historic journey:

> There was great cheering and Rock Cannon firing all along the line—and on our arrival at Port Madoc over *the* Embankment, we were drawn by horses . . . at Port Madoc we were met by Crowds of People, Bands playing—and the Workmen had a good dinner given them—all the better Company

were entertained at Morfa Lodge by James Spooner and ended the day with a dance.[9]

Samuel Holland used the railway to carry slate to Portmadoc for almost two years before the other companies followed his example; and he rented a wharf at the harbour until 1877. At first the railway used to climb a ridge of hills; but in 1840 the Moelwyn tunnel was opened. This allowed the trains to run by gravity all the way from Blaenau Ffestiniog to the embankment. Horses were used to pull the empty wagons back up again; and on the down run they travelled at the end of the train in special horse boxes, usually feeding all the way. The average gradient of the line was 1 in 92, and there were many short curves. In January 1865 the line was opened for passenger traffic; and it proved of inestimable value, not only for carrying the slates to Portmadoc but also for transporting the workmen to and from the quarries. Goods including foodstuffs were imported through Portmadoc and distributed throughout the developing region; and Portmadoc became the only port for the Ffestiniog quarries.

Before the opening of the Festiniog Railway, the carriers took their slates to Traeth Bach and then proceeded to Tremadoc to buy provisions. The railway provided a cheaper and quicker method of transport: ships from Liverpool brought goods daily, and these were then sent out to the surrounding district by rail. As the demand for ships increased, shipbuilding became a significant industry in Portmadoc; and it was also developed in Aberdovey, Barmouth, Pwllheli, Nevin, Porthdinllaen, Port Dinorwic and Bangor.[10]

Portmadoc increased rapidly in commercial importance; and the population of the parish of Ynys Cynhaiarn, in which Portmadoc was situated, rose steadily from 1801 to 1881:

Year	*Population*
1801	525
1811	889
1821	885
1831	1,075

Year	Population
1841	1,888
1851	2,347
1861	3,138
1871	4,367
1881	5,506

Until the 1850s this increase in population was the result rather of natural increase than of immigration; but from 1851 to 1881 there was a considerable movement of people into Portmadoc.[11]

The tonnage of slates carried across the embankment from the Ffestiniog quarries to Portmadoc harbour increased from 4,275 in 1836 to 120,426 in 1882, the peak year. This large-scale development of the slate industry in North Wales was brought about by organisations of various types. Some were financed by wealthy capitalists; but many were common law partnerships of preachers, architects and quarrymen. In the Nantlle area the Dorothea Quarry became the largest and most profitable concern. This was at first known as Cloddfa Turner, and belonged to a family firm established about 1829 by William Turner and his son-in-law John Morgan of Caernarvon. The quarry was situated about nine miles from Caernarvon, on the north side of the lower Nantlle lake in the Vale of Nantlle; and it extended for half a mile along the course of a slate bed. The profit was estimated at about £2,000 per annum; but as William Turner grew older much of the management was entrusted to his son Thomas and to the local manager O. Parry, and the profits began to fall. The landlord was Richard Garnons (1774–1841); and the name Dorothea seems to have been given to the quarry as a compliment to his wife, whose Christian name it was.

The first pit to be worked was Hen Dwll (Old Pit); and primitive machinery consisting of turntrees and pumps worked by hand, was used there to lift slate blocks and debris. The slates were blue and often contained stripes; this may be why the pit was closed in the 1840s and allowed to flood. The second pit was known as 'Twllyweirglodd' (Meadow Pit); the rock there was at first excellent, but as it got deeper, the slate vein became more stripy. In 1841 a third pit was opened, whimseys being used to clear it. These whimseys having proved inefficient, a steam engine was erected and as the men called this engine a 'fire engine', the pit was named 'Twll Fire'. After many years this pit and Twllyweirglodd became one. 'Twll Coch' (Red Pit) was opened about 1843, and had two steam engines for raising blocks and debris. The slates there were purple in colour.

Turner and Morgan decided to close Dorothea Quarry in April 1848, and it was put up for sale. They issued a circular, which stated that the quarry and premises were held under lease for the unexpired term of 22 years. The minimum rent was £250 per annum, and the royalty was 3s per ton on all slate brought down to Caernarvon. The quarry was served by the Nantlle Railway Company, which charged tolls of 2s 11½d per ton slate, 8¾d per ton on coal, 3s 3d per ton on iron and 1s 7½d per ton on wood. The quarry was said to employ 200 men and boys; and the output was given as 5,000–6,000 tons per annum. The quarry possessed 'Engines, pumps, weighing machines, sheds, railroads and waggons'; and it was said that the quarry was an 'eligible investment' with 'a large connection demanding more than the present supply of slates'.

A small group of Nantlle quarrymen, including William Owen, John Jones, and John Robinson, had the idea of working the quarry as a company; and they received encouragement in this from the Rev John Jones of Talysarn, who had been a quarryman. The quarry was bought by them for £3,000; and 100 shares were issued at £25 each, although 200 shares had actually been authorised. William Owen of Hafodlas was the first manager of the workman's company; but as a result of a disagreement, he was succeeded in June 1851 by the Rev John Jones, who managed the concern on a temporary basis, for about a year. John Robinson was appointed to succeed him, and on 30 June 1857 a new lease of the Dorothea Quarry was granted by John W. M. Gwynne Hughes to John Lloyd Jones, John Robinson and Thomas Lloyd Jones for thirty-five years. The royalty varied according to the output and quality of the slates: it was 2s 6d per ton, but on striped slates above 5,000 tons it was only 1s 6d. The dead rent was £300 per annum.

By this date the original idea of a working men's partnership had disappeared. In the early years this had been carried out, with meetings held every two months at the quarry office; but the quarrymen later realised that their holding of twenty shares was insufficient to give them a controlling voice. In September 1853 they sold their shares to John Williams of Plas yn Blaenau in Denbighshire for £100 each, which was four times their original value. Williams had not been one of the original shareholders; but he had purchased four shares in May 1853. In April 1858 he bought another two shares at £150 each; so that in all he had then acquired twenty-six shares for £2,700. The Rev John Jones died on 16 August 1857; and on 12 April 1860 Williams bought out one of the lessees, John Lloyd Jones, paying £250 for each of his seven shares. In August 1860 the company resolved to transfer its banking account from Williams & Company to the North & South Wales Bank in Caernarvon. A lease was granted in March 1862 by Gwynne Hughes to John Williams, John Robinson and Thomas Lloyd Jones for thirty-one years; the terms were similar to those of the 1857 lease.

In 1864 attempts were made to sell Dorothea Quarry; and a survey of the quarry was made in August of that year. The consequent report described a total of sixty slate-producing bargains in the quarries, which yielded nearly 1,000 tons of slates per month. It declared that a large amount of capital had been spent to bring the quarries to their flourishing condition, and went on to enumerate the machinery and plant. The list included:

> Three Fire Engines, nearly new, with Drums, Pulleys, Chains and all connecting gear for the purpose of hauling up the produce, three large water wheels for pumping out the water and working sawing and planeing machines, two sawing and one planeing machine, Tramways, Trucks and Waggons, Offices, weighing machines, dressing sheds, Smiths and Carpenters Shops and various other buildings and cottages erected for the accommodation of the workmen.

The report pointed out that the quarries had a direct tramway to Caernarvon, by which they could convey goods cheaply with a great saving in breakages.

The negotiations for the sale of Dorothea were unsuccessful; and John Williams continued to buy shares in the concern. In November and December 1864 he bought 4 from Thomas Lloyd Jones and 2 from John Robinson at £300 each. John J. Evans from Cilgwyn was a manager of Dorothea from 1864 until 1874, when he went as manager to Penrhyn Quarry. Under his management the men were contented and profits were high. John Williams died in March 1879, by which time he had acquired 71½ shares out of the 100 shares. Three people were appointed to take control until the coming of age of I. H. G. Williams in May 1890.[12] The family continued to hold a majority of shares in the quarry until it was closed in 1970.

The Company Act of 1862 did much to encourage investment in the slate industry. By subscribing a memorandum of association, any seven or more partners could form a company either with or without limited liability. This removed the risk to private capital which had hitherto existed. The late Dylan Pritchard estimated that by 1882 thirty-one of the sixty-five slate-producing units in North Wales were joint stock limited liability companies. These thirty-one companies employed more than one-third of the area's total number of quarrymen; but they produced less than one-third of the area's total slate output.[13]

The Dorothea Company became very profitable, and from 1849 to 1859 the total value of the slates sold rose from £5,427 to £16,630. In the years 1852–9 the following dividends were paid: 1852, £400; 1853, £500; 1854, £1,000; 1855, £750; 1856, £500; 1857, £800; 1858, £1,000; 1859, £650; 1860, £856. Higher profits were achieved during the period 1861–82. In the years 1861, 1863, 1864 and 1879 losses of £291, £321, £114 and £2,062 were made; but in most other years there were substantial profits and dividends:

Year	*Profit*	*Dividends*
1861 (½ year)	£34	
1862	£2,639	
1863 (½ year)	£778	£500
1864 (½ year)	£908	
1865	£4,234	£2,000
1866	£4,010	£2,000
1867	£4,201	£2,000
1868	£9,154	£5,000
1869	£8,413	£6,000
1870	£4,678	£6,000
1871	£4,485	£6,000
1872	£5,560	£6,000
1873	£8,769	£6,000
1874	£10,553	£8,000
1875	£14,738	£10,000
1876	£10,714	£16,000
1877	£10,882	£12,000
1878	£11,439	£6,000
1879 (½ year)	£3,023	
1880	£6,149	
1881	£6,393	£6,000
1882	£4,585	£8,000

It will be seen that profits reached their peak in 1875, and that the highest dividend was that for 1876. The output figures for the years 1860–82 were as follows:

Year	*Tons*
1860	3,497
1861	8,434
1862	9,740
1863	10,343
1864	9,218
1865	10,312
1866	9,832
1867	10,501

Year	*Tons*
1868	13,687
1869	15,857
1870	13,613
1871	12,399
1872	15,910
1873	17,422
1874	15,607
1875	14,860
1876	12,561
1877	12,553
1878	11,188
1879	12,097
1880	13,080
1881	16,853
1882	14,197

It is clear that the company's most prosperous years were around 1873–6.[14]

The Dorothea Company was the largest and most profitable concern in the Nantlle Valley. No quarry there was on the scale of Penrhyn and Dinorwic, but several were large, and all had benefited from the improved transport provided by the Nantlle Railway. In 1873 the Dorothea was said to be 'by far the most extensive quarry in this district'; at that date it employed 450–500 men. The co-partnership running the Cilgwyn Quarry was dissolved in 1830;[15] and the quarry was then worked by an MP named Muskett. Muskett's period in control came to an end when he fled from the country, leaving debts of £10,000; and he he died in 1844. During the 1840s the quarrymen whose wages had not been paid, worked the quarry on their own account and sold their slates at very low prices. They concealed their digging as far as possible; but some of them were sent to prison two years later. About 1846 a new company took over the quarry with an Oswestry solicitor named John Hayward as the principal proprietor; Hayward was succeeded in 1849 by his son William. The four pits of the quarry were Faengoch, Old Cilgwyn, Cloddfa'r

Dwr and Cloddfa Clytiau; and in 1858 9,000 tons of slabs and roofing slates were produced. In 1873 Faengoch and Cloddfa Clytiau were the only pits being worked.

The method of working these pits was very hazardous. The pit was reached by a wagon, which traversed two steel wires leading to the bottom of the quarry. The steel wires were laid almost vertically and the wagons were attached to a girder. In 1870 a large fall of rock occurred in Faengoch, covering the floor of the pit. In 1873 the number of men employed was 260.[16] In January 1880 a letter from J. Aussell Sowray of the Office of Woods reported that the workings at the Cilgwyn Quarries were unworthy of the property's 'character, value and productiveness'. The slates at Faengoch had been obtained from blocks picked out in clearing the fall which had taken place in October 1878; but 'had a moderate sum been laid out annually in clearing the tops round this Quarry, the interruption caused by the last fall would have been avoided and the future security of the Quarry ensured.' At Old Cilgwyn the workings were 'almost abandoned'; and at Cloddfa Clytiau the lower galleries had been worked more rapidly than the upper ones, 'rendering falls probable during the winter months'. In a marginal note, William Hayward confirmed that the quarry was 'sadly deteriorated as to its value and productiveness in consequence of the depth of Granite over the slate rock'. Other factors operating against the quarry were the lack of tipping space and the fact that Cilgwyn slates were often rough and striped.[17]

A little to the north of Dorothea Quarry was Penybryn Quarry, known also as Cloddfa'r Lon and later taken over by the Dorothea Company. In 1873 the quarry consisted of 4 pits, and as they were near the Nantlle lakes there was always a danger of flooding. The quarry had been opened in the 1770s when a few local workmen dug for slates. It had been taken over in 1808 by William Turner, Hugh Jones, William Wynne and Richard Garnons; but a series of disagreements caused it to be given up. Richard Garnons, the landlord, worked it himself till 1834; and then a new company acquired it. The number of men employed in 1873 was 300.

The smaller quarries in Nantlle included one owned by the Braich Slate Company, which was formed about 1869; this was in the parish of Llandwrog, and in 1873 it employed 140 men. The quarry was held from the Crown, and had been worked unsuccessfully about 1820. It consisted of one large pit, which had three floors. Nearby was the Fron or Vron Quarry, which employed seventy men in 1873, and consisted of one pit. It was owned by a limited liability company which owned the pits collectively known as the Talysarn Slate Quarry. The slates were transported by the Talysarn Railway, which joined the North Wales Narrow Gauge Railway. This in turn joined the London & North Western Railway at Dinas Junction.

The Galltyfedw Quarry, one of the oldest in the valley, was situated between Dorothea and Cilgwyn; it consisted of two pits, one of which was flooded by 1873. At that date the quarry had about twelve men working it; and it was owned by R. D. Williams of Caernarvon and a company of 'private gentlemen'. At the bottom of the Nantlle Valley there was another early quarry known as Cloddfa'r Coed Quarry; in 1873 it was owned by H. Roberts, and three or four men were engaged in quarrying the fine blue-grey slates. This quarry was later taken over by Thomas Robinson, who also owned the Talysarn, Tan'rallt, Blaen y cae and Braichmelyn Slate Quarries.[18]

Penyrorsedd Quarry, another Nantlle concern, was opened in 1816 by William Turner; and in 1854 John Lloyd Jones acquired possession of it. In 1862 it was taken over by the Darbishire Company, in which William A. Darbishire of Lancashire was the principal shareholder. Over £20,000 was unwisely expended in sinking shafts and driving tunnels; and about 400 men were employed. The resulting output was only 2,000 tons; and by 1871 this had decreased to 500 tons, with the number of men reduced to forty. At this point Darbishire took a more active part; and under his direction the quarry flourished. In 1882 the output was 7,999 tons and 260 men were employed. The quarry was worked in galleries, and the slates were manufactured in covered sheds.[19]

Many of the smaller Nantlle Quarries failed to prosper, often because of wasteful capital expenditure; and there was also a very great scarcity of tipping ground, especially at the bottom of the valley. Another quarry affected by unwise and uncontrolled capital expenditure was the Gorseddau Quarry, which lay close to Llyn Cwmystradllyn in the parish of Penmorfa about two miles south-west of Beddgelert. A company was formed in 1854 to work the quarry; but instead of making preliminary trials to establish the nature and extent of the slate rock they spent £52,000 on capital equipment. The slate vein of this quarry, which was on the slopes of Moel Hebog, proved in fact to be narrow and of very poor quality. On 25 May 1855 the quarry was opened by R. Gill and J. Harris for the Bangor and Portmadoc Company; and around the same time that company developed plans for a tramway from the quarry to Portmadoc.

This tramway was designed to carry slates to Portmadoc from the quarries around Snowdon and Moel Hebog; and in 1855 O. Griffiths agreed to sell his land for this purpose at £100 per acre. A horse tramway was constructed under the authority not of an Act of Parliament but of wayleaves granted by the landowners. The line used part of the course of the Tremadoc Tramway, which connected an ironstone mine at Llidiart Yspytty, west of the town, with Portmadoc harbour. It ran from the quarry to Tyddyn-mawr, and was said to have cost £16,000. By an Act of July 1872 (35 and 36 Vic cap 155) the Gorseddau Junction & Portmadoc Railway Company was authorised to build a five-mile extension to the Prince of Wales Quarry at the head of the Pennant Valley. The Act authorised not only locomotive traction but also passenger traffic; but when the line was opened after reconstruction in September 1875 it handled goods traffic only. The gauge of the original line was 3ft; but this was subsequently altered to 1ft 11½in, which was the gauge of the Festiniog Railway. The rolling stock in 1878 consisted of one locomotive and fifteen wagons for slates and other goods. In 1892, however, the Gorseddau crossing of the Cambrian Railways to the east of Portmadoc was discontinued; and by 1894 the line had ceased working. Part of the route can still be seen on the ground.

The other industrial monument left by the concern is a large slate factory or mill, now gaunt and roofless, which once contained expensive machinery. The mill lies between 1½ and 2 miles from the quarry, which was the nearest point for obtaining sufficient water-power. It was fed water from a stream, which had a weir built across it. The water was brought by an aqueduct to the south-eastern end of the mill, where it entered a tunnel and finally reached a waterwheel. (No remains of this wheel have been recorded.) The mill was a three-storeyed building, resembling an early textile mill rather than the usual one-storeyed slate mill; and two storeys were built into the steep slope of the river bank. On the top floor the slate was sawn; on the middle floor the slate was dressed; and from the ground floor the finished slate was carried by railway to Portmadoc. Financially, the company was a disaster. The output of slates in 1859 was only 1,338 tons, although over 200 men were employed. In 1862 only 90 tons were produced. In 1867 the company was wound up, and in 1870 it went into liquidation. In 1872 the Prince of Wales Company purchased the quarry for £5,000; and they were responsible for linking the tramway to their quarry. They hoped thereby to reduce their heavy transport costs, as it cost about 12s 6d a ton to carry slates by cart to Rhyd-Ddu and then to Caernarvon. The prospects of the Prince of Wales Quarry seemed good; and the quarry consisted of seven galleries and a large machine room. In 1873 it was a limited liability company employing 200 men and producing 5,000 tons of slate annually. It was ruined by a large fall of rock, which put a stop to slate production; and in 1882 there were just a few men working to remove the rubble.[20]

Pennant Vale, 'an exceptionally bleak and barren place between Beddgelert and Criccieth', contained a number of quarries; but by 1882 all but one, the Prince Llewelyn Quarry, had ceased to operate. The slate quarries around Snowdon included that of the South Snowdon Company, which in 1878 sent a wagon to Portmadoc every day 'with a fine team of horses'. At the same date the Snowdon Slate Quarry on Bwlch Cwmllan produced about seventy tons a month. The Moelfra Quarry, which was situated on the same vein as the Prince Llewelyn Quarry, employed between twenty and thirty men in 1873.[21]

In the Eden Valley the slate resources were not well developed because of poor transport facilities. Until 1879, when the London & North Western Railway Company constructed the line between Bettws y Coed and Ffestiniog, slate from the Dolwyddelan quarries had to be carted to Bettws y Coed; and after 1876, when there was a fall in slate prices, many of the smaller concerns had to close because they could not compete. In 1873 there were four quarries in the Dolwyddelan area: Gethin Quarry above Dolwyddelan, opened in 1870 by Owen Gethin Jones of Penmachno; Penllyn Quarry; Pompren Quarry; and Prince Llewelyn Quarry.[22]

The slate resources of the Beddgelert or Gwynant Valley were developed after 1840; but the quarries were all closed by 1882. Henry McKellar was a director of Hafod y Llan Quarry or South Snowdon Slate Quarry, and Sygun Copper Mines at the head of the Gwynant Valley. The quality of the slate was poor, the proportion of waste to slate was high, and the expense of carrying the slates to Portmadoc was another serious disadvantage. Allen Searell, a native of Devon, came to the Caernarvon-Merioneth border in order to develop mines and quarries; and he worked conscientiously as manager for his London directors, of whom McKellar was one. Some of his correspondence with McKellar has been preserved in the Searell MSS in the University College of North Wales. In 1844 he went as manager to Cwmorthin Quarry; but he returned to Hafod y Llan some time before his death in 1865.

In 1843 a tramway was projected from the Snowdon slate quarries to Portmadoc; but as the Hafod y Llan Quarry was then in a very early state of development, with only nine to twelve men working, little progress was made. The scheme was discussed again in the 1860s; but it was never carried out, chiefly because of the difficulty of persuading landowners to sell their land. About 1860 the cost of this proposed railway from the Snowdon Quarries to Portmadoc was estimated at £13,960 10s 6d. This price was independent of compensation for land. The regulations of Hafod y Llan Quarry included one whereby every group of men taking a contract would be required to work ten hours on this, and failure to do so meant a fine of 2s 6d. Other fines included 5s od for being drunk at the quarries; and 5s od for the illegal removal of slate that had been manufactured and checked. All those employed in the quarries had to pay 2d in the pound of their actual earnings, after deductions, to the doctor 'for attendance and medicine'.[23]

In the Conway Valley, capital was attracted to slate workings in the 1860s. Lord Newborough of Glynllifon leased Cedryn Quarry to Richard Griffith of Bangor, who was a banker, and William A. Darbishire, the chief shareholder of the Penyrorsedd Quarry. A twenty-one year lease was taken out in December 1862; and the lessees formed a limited liability company with an authorised capital of £50,000. A railway was constructed from the quarry down to the River Conway near Borthllwyd.[24] Cwm Eigia, another remote quarry in the Conway Valley, was leased from Sir R. B. W. Bulkeley by the British Slate Company Ltd, which had an authorised capital of £30,000. The total production from 1 July to 31 December 1865 was 840 tons or 367,900 slates. In the first half of 1866, 448 tons of slate were produced, and 342 tons of slate in the second half of 1866. In 1867 the quarry produced 588 tons of roofing slates and 140 tons of slabs. The royalty on slabs was one-sixteenth of the value; and that on roofing slate was 2s per ton.[25] These quarries were short-lived, partly because their slates were of poor quality and partly because their transport costs were high.

The quarries of Corris were situated on the Llandeilo Range, which extended from Dinas Mawddwy to Towyn. The main quarries were opened on a pass leading from Talyllyn; and upon the southerly side of the pass were Gaewern and Braich Goch Quarries. A forty-year lease of Gaewern Quarry, five miles from Machynlleth, was obtained in 1857. In the quarry there were two sections of the slate vein, one running parallel to the pass from Talyllyn and the other going southwards towards Bryneglwys Quarry. The quarry was situated on the side of a hill and was worked mainly by mining. The slates were transported by the Corris, Machynlleth & River Dovey Tramroad, which was authorised by Parliament on 12 July 1858 and opened for goods traffic on 30 April 1859. This tramroad was used by quarries in the Corris neighbourhood, which had previously relied on carts and sledges to bring their slates to the river; it had a gauge of 2ft 3in, and was later called the Corris Railway. The Act of 1858 authorised three lines: Machynlleth to Aberllefenni; Braich Goch to Upper Corris; and Aberllefenni to Ratgoed. In 1873 the number of men employed was 200; and on 21 March 1882 this 'old established quarry' was put up for auction. The advertisement claimed that the quarry had been established for sixty years, and that it produced 'the finest Slates and Slabs'. The property for sale included the manager's house and twenty-eight cottages, besides engine houses and sheds, which were connected with the workings by inclines or tramways. The quarry was purchased in February 1884 by the Braich Goch Company.[26]

Braich Goch (or Braichgoch) was near the village of Corris, and was worked in underground galleries. It produced slabs as well as roofing slates, and employed 200 men. The capital of the Braich Goch Company was £51,000, and the profits from 1879 to 1882 were as follows:

£4,610 on 31 March 1879
£586 ,, ,, ,, 1880
£617 ,, ,, ,, 1881
£291 ,, ,, ,, 1882 [27]

The other quarries in the Corris area included Cambria Wynne Quarry, whose output consisted almost entirely of slabs. This was a manufacture which had developed in the 1860s and provided an additional source of income for the slate quarries. The slabs were made from pieces of rock which had been hardened by heat, or were crossed by hard bands, or lacked the silica necessary for good cleavage. They were planed, and were used for a variety of purposes: for the flooring of barns and farm buildings generally; for the flooring of houses; for doorsteps and sills; for public urinals; for mantelpieces; for stalls and feeding troughs; for water-cisterns; for pavements; and for tombstones. The preparation of slabs for these purposes was usually separate from the quarry undertakings. In and near Bangor, Caernarvon, Portmadoc and Llangollen there were factories for this purpose;

and they obtained the squared and planed slabs from the quarries. At Llangollen the slate mill was worked by the quarry owners.[28]

The Aberllefenni Quarry, which lay to the north of Corris, had been worked as far back as the sixteenth century; but it was not developed on a capitalist basis until 1810. It consisted of three sections, which were known as Hen Chwarel, Ceunant Du and Foel Grochan. Hen Chwarel was a large open pit, and employed about 180 men. Foel Grochan produced only slabs, as the rock was too hard to be made into roofing slates. The three sections were worked by the same proprietor. North of Aberllefenni were three small quarries served by the Corris Railway: Cwm Era, Hendre Ddu and Ratgoed.[29]

Bryneglwys (or Bryn Eglwys) Quarry at Abergynolwyn was connected to Towyn by a narrow-gauge railway known as the Talyllyn. The Talyllyn Railway was authorised by Parliament in 1865; and the line, which had a gauge of 2ft 3in, ran from a junction with the Aberystwyth & Welsh Coast Railway at Towyn to Abergynolwyn. From Abergynolwyn a mineral extension ran to Bryneglwys Slate Quarry. In 1873 the village of Abergynolwyn was composed entirely of houses built for the quarrymen; and it was said to have a 'new and clean appearance'. The quarry was first worked by a group of workers, who were reputed to have sold it to John Pugh in 1847 for £120. The quarry was on the Hendre estate, which was owned by Richard Pugh and after 1858 by his son. In 1864 William and Thomas Houldsworth McConnel, Manchester cotton manufacturers, formed the Aberdovey Company; and this company, which had a capital of £75,000, leased land on Hendre estate to work the slate. They promoted the Talyllyn Railway, which was opened in December 1866. The number of workmen in 1873 was 260; most of them lived in Abergynolwyn, but a few houses had been built high in the hills near the quarry. There was a barracks for those who lived at a distance, and a Sunday School was held in its kitchen. There was a writing slate factory at Abergynolwyn. The company produced about 500 tons of slate a month; but the investment had to be increased to over £100,000, and the output was not sufficient to pay dividends. In 1883 the company went into liquidation; and all its assets, including all shares in the Talyllyn Railway, were purchased by the McConnel brothers.[30]

Dinas Mawddwy Quarry was situated on the slopes of a hill called Bryn yr Wylfa, and was mainly important for its output of slabs. Blocks were conveyed down inclines to a large machine-room where there were over forty planing and sawing machines driven by water power. The quarry had a standard-gauge railway 6¾ miles long; and this was owned by a private company whose principal shareholder was Sir Edmund Buckley. The track, which ran from Cemmes Road to Dinas Mawddwy, was opened on 1 October 1867. At first it depended almost entirely on the slate traffic of the Dinas Mawddwy Quarry; but connections were soon made with the other slate quarries in the Upper Dovey Valley. The railway was later linked with the Cambrian Railways at Cemmes Road.[31]

In north Montgomeryshire there was an old slate-producing district on the Llandeilo strata near Llangynog; the principal quarry, the Rhiwarth, had been worked since the early eighteenth century. Quarries worked on the Wenlock strata in 1873 included Clettwr near Llanderfel; Penarth near Corwen; Oernant and Cloggau near Llangollen; the Glyndyfrdwy Quarry of the Llangollen Slate and Slab Company; Moel Ferna in Glyn Ceiriog; the Cambrian near Glyn Ceiriog; and the Wynne near Glyn Ceiriog. The heavy cost of transport limited the development of these quarries.[32]

The Nantgwryd Quarry in Glyn Ceiriog was being worked at the end of the eighteenth century, when it belonged to Richard Myddleton of Chirk Castle. In the 1870s a new undertaking was working the quarry; and it was then hoped that the 'almost unknown and inaccessible region lying between Bettws y Coed and Llanwrst, and the Vale of Clwyd' would shortly develop its slate and slab quarries. These small quarries played little part in the economic development of the slate industry; but some of them contributed to an improvement in conditions for the local people. In August 1905 a group of workmen recalled the consequences which had followed the opening of Braich Goch Quarry:

> Before the arrival of your company Corris, at its best, was but a poor and unsatisfactory place to live in. Everything was there in an undeveloped state—work was difficult to obtain, wages were poor and indifferently paid, and the number of inhabitants but small in comparison to what it is now. The advent of your Company at once made a great change. A new era of prosperity set in for Corris. Houses were built and the people flocked to the place, with the certainty that plenty of work would be found for capable and willing hands to do.[33]

The most successful quarries in southern Merionethshire were those with access to the Corris Railway, which was converted to steam power in 1879. Two of these, the Aberllefenni and Braich Goch Quarries, are still working; but since the closing of the railway in 1948 they have used road transport.[34]

The slate villages of the Ffestiniog area included Fourcrosses, Congl y wal, Rhiwbryfdir, Tangrisiau and Ffestiniog itself. A journalist touring the area in 1873 described the growing town of Blaenau Ffestiniog as the 'City of Slates'; the walls, parapets and kerbstones were made of slate, and in the houses pieces of polished slate were used as mantelpieces and tabletops. He declared that the mud on the Ffestiniog roads was of a 'blue slaty colour'; and he was enchanted by the picturesque view from Ffestiniog village, with the mountains in the background and the River Dwyryd winding like a silver thread along the valley. Outside the crescent of villages were the quarries: Cwmorthin, which was later owned by the New Welsh Slate Company; Croesor; Rhiwbryfdir; Holland's; Palmerston, which was owned by the Welsh Slate Company; Llechwedd; Rhosydd; Rhiwfachno; Diphwys Casson; Votty and Bowydd or Lord; Maenofferen; Cwt y Bugail; Rhiwbach; Blaen y Cwm; Bwlch Slaters; Craig Ddu; Drum; and Foelgron. Rhiwfachno and Rhiwbach were in Manod Valley not far from Dolwyddelan.[35]

During the period 1860–74 Croesor Quarry was run by the Croesor United Slate Company, which had a capital of £160,000. The Croesor Tramway, which had a gauge of 1ft 11½in, was

Page 137 (*above*) Llechwedd Quarry on pay day about 1890, showing the haulage house, the waterwheel for haulage, floor 4, and the steam haulage house (with chimney) on floor 5; (*below*) Craig Ddu quarrymen returning home down the incline on the trolleys or 'wild cars', about 1900

BLAENAU FESTINIOG. GRAIGDDU QUARRYMEN RETURNING HOME DOWN INCLINE ON TROLLEY

Page 138 (*above*) Dinorwic Quarry, with a view of the velocipedes, about 1892; (*below*) the Nantlle Railway with the trucks outside the Penyrorsedd stables

constructed by Hugh Beaver Roberts about 1863 without parliamentary authority. It ran from Portmadoc to the upper end of the Croesor Valley, a distance of about seven miles; and in October 1863 a deed of Mutual Covenants was made between Roberts and the Rhosydd Slate Company Ltd, by which a wayleave was granted enabling the Rhosydd Company to construct a railway line and incline from its property to the Croesor Tramway. In 1867 the Croesor Quarry produced 3,221 tons of slate; but there was a loss of £498. In the following half-year the output was 1,906 tons and a profit of £321 was made. This information was contained in a circular issued for the benefit of the shareholders; and notes in ink on one extant copy suggest that the report may have been too optimistic. The circular stated that nine chambers were working at a profit, that eight of these had an average life of three years, and that two chambers were nearly unroofed; but these statements are queried in the MS notes. James Wyatt, Lord Penrhyn's former agent, was director of the company until 1866, and he gave a detailed account of its expenditure on 18 February 1862. Apart from the construction of the railway and the wharf, the main items of capital expenditure had been the opening of chambers and the preparations for slate-making. The output was estimated at 600 tons per chamber per annum, and the profit at around 21s per ton. In 1867 there were 175 men employed, but in June 1868 the number had been reduced to 111.

The Atlas Assurance Fire Policy, taken out on 2 April 1869, insured a slate-sawing mill, together with the engine house under the same roof, for £400; a waterwheel and machinery for £700; a building containing a ventilating machine for £150; carpenters' and smiths' workshops and store rooms for £150; a barracks and an adjoining cottage some distance from the workshops for £300; a stable, part of which was sometimes used as a barracks, for £100; a house in the valley which housed two quarrymen for £300; and another house in the valley for £300. The total value of the insured property was £2,550.

An interesting comparison of Penrhyn and Croesor Quarries was made in 1869, in an attempt to show why Croesor was a losing concern and not a profitable one like Penrhyn. The comparison, made in terms of output and costs, ran as follows:

	Penrhyn	*Croesor*
	Tons Cwt	Tons Cwt
Product of slate, men and boys	5 0 per head	5 8 per head
Gross value of slates	42s per ton	55s per ton
Value less discount	42 " "	50s " "
Cost of getting and making including bad rock	16s 7d per ton	25s 3d per ton
Wages of Slate Men and boys	£4 3s 2d	£6 10s 1d
Wages of Slate Men and Boys clear of deduction	£3 19s 2d	£5 17s 0d
Gross wages of the whole quarry	£4 1s 0d	£5 17s 0d
Net wages of the whole quarry	£3 17s 0d	£5 7s 4d
Fixed charges viz salaries and management	1s 2½d per ton	10s 9½d per ton
Interest and wharf Rent	£3,000	£110

Notes added at the bottom of this comparison explained that the fixed charges on Croesor were necessarily high, because the bargains were few and produced only 226 tons monthly compared with almost 9,000 tons at Penrhyn. Thirty-two chambers had been opened at Croesor; but only seven of these had turned out well. Since its establishment, the quarry had produced 7,873 tons of slate. The chief handicap at Croesor, according to this account, was the high cost of working the quarry; and this resulted chiefly from high wages and heavy premiums.[36]

One of the oldest quarries in the Ffestiniog district was Diphwys Casson, which lay 985ft above sea-level. It was worked by William Turner, William Casson and Thomas Casson; and they eventually acquired the freehold. The slate vein at Diphwys Casson was close to the surface; and this reduced costs. In 1863 the quarry was sold for £120,000 to a limited liability company; and in 1873 the number of men employed was 279. The quarry was worked in terraces along the sides of Diphwys Hill but underground levels were also worked.[37] Rhosydd Quarry was situated on Moelwyn Mountain, 1,495ft above sea-level; and it was worked by William Turner and by various small companies. In 1857 it was bought by a London company called the Rhosydd Slate Quarry Company Ltd, and the problem of getting slate down from the mountain-side was solved in 1866, when the quarry was linked by inclines to the Croesor Tramway. In 1874 a number of local gentlemen formed another private company, which had a nominal capital of £80,000. At that date 83 men only were employed; but five years later this had increased to 181.[38]

Lord or Votty and Bowydd Quarries, 840ft above sea-level, were first worked by Lord Newborough at the beginning of the eighteenth century; but little capital was expended on them until in 1834 Bowydd Quarry was leased to Edward Shelton and J. W. Greaves. Greaves bought this quarry after Shelton's death; but he thought that the slate vein was exhausted, and on the termination of the lease turned his attention to Llechwedd. Near Bowydd Quarry there was another working, also on Lord Newborough's estate. This was close to a farm called Hafody, and the quarry was later known as Votty. A geological fault separated the quarries, so they had to be worked independently; but because of the steep dip of the slate beds, underground working was necessary in both. After 1846 the two quarries were acquired by a limited liability company, in which the principal shareholder was F. S. Percival of Northampton. In 1873 the number of men employed was 310. The whinstone underlying the old vein was removed, revealing the 'real blue vein' below the lower land. In December 1874 the nominal capital was £80,000; and in the same year the value of the 'Quarries, Plant and Machinery' was given as £61,000.

In June 1875 the company paid £408 18s in rents and royalties to Lord Newborough. After the half-year ending on 3 June 1875, the company declared an interim dividend of 6 per cent on the paid-up capital of £66,000; and dividends continued to be paid at the following rates:

½ year ending 30 June	1876	8 per cent
½ year ending 31 December	1876	8 ,, ,,
½ year ending 30 June	1878	10 ,, ,,

½ year ending 28 June	1879	6 per cent
½ year ending 3 January	1880	6 ,, ,,
½ year ending 26 June	1880	3 ,, ,,
½ year ending 8 January	1881	4 ,, ,,
½ year ending 7 January	1882	1½ ,, ,, [39]

By the 1870s the Welsh Slate Company had excavated an enormous amphitheatre on the slopes of Moelwyn; and the two other Rhiwbryfdir Quarries, Middle Quarry and Holland's Quarry, were attacking the same mountain. The Welsh Slate Company had acquired its quarry, then called Rhiw or Rhiwbryfdir, from Samuel Holland and his father. In 1842 it had debts of £60,000; but under the chairmanship of Lord Palmerston, who used to pay an annual visit to the quarry, it became extremely successful. In 1873 its quarry was the biggest in Ffestiniog; with George C. Chissel as manager, it employed about 700 men.

The quarry consisted of large chambers 50–60yd deep and about 20yd wide; and these alternated with pillars of comparable width, which supported the roof.[40] In 1883 parts of the quarry collapsed, causing extensive damage to the Middle Quarry. Damages of £110,000 were awarded to the Oakeley Company; and the Welsh Slate Company therefore surrendered the remainder of the lease to them.

Holland's Quarry employed about 500 men in 1873; and by that date Samuel Holland had introduced gas lighting in his underground works. In 1878 W. E. Oakeley took over this quarry, together with the Middle Quarry or Matthews' Quarry. Middle Quarry was so called because it was between Holland's Quarry and that of the Welsh Slate Company. It was worked partly in open levels and partly in underground tunnels. It was worked from 1838 to 1878 by Nathaniel Matthew; and in 1874 it employed 416 men.[41]

In the Rhiwbryfdir district also was Llechwedd Quarry, 780ft above sea-level and destined to become one of the largest undertakings of Ffestiniog. This quarry was opened in 1846 by John Whitehead Greaves, his son A. Ernest Greaves being the manager. In place of the guillotine, which had been operated rather jerkily by the revolution of a wheel, J. W. Greaves invented a slate-dressing machine with a circular action; this dressing machine was in use by 1873. At that date the underground workings at Llechwedd consisted of vast, cathedral-like chambers and the number of men employed was 330. The quarry's output increased from 2,895 tons in 1850 to 19,958 tons in 1880.[42]

The highest quarry in North Wales was Craig Ddu, which was opened in 1840 and lay 1,800ft above sea-level. A tramway was laid from the summit to the road in a series of inclined planes; and the men descended the inclines in small linked cars, each of which had a front wheel designed to fit one rail and a bar with a roller extending to another. These vehicles were known as 'ceir gwyllt' or 'wild cars'. The quarry was worked in the open; and in 1873 it employed 106 men.

The Bugail and Cwt y Bugail Quarries were between Penmachno and Blaenau Ffestiniog; and reaching them from either place involved a steep four-mile climb. Slates were sent to the Festiniog Railway by a tramroad from Rhiwbach Quarry. The works at Bugail and Cwt y Bugail were partly underground and partly open; and in 1873 over a hundred men were employed, most of them living in barracks during the week and returning to their homes at weekends.[43]

North of Blaenau Ffestiniog and above Votty and Bowydd Quarries was Maenofferen Quarry, 1,035ft above sea-level. It had been worked spasmodically in the early nineteenth century, but it was not until 1861 that the quarry was worked successfully. The new limited liability company, which leased the quarry from Lord Newborough, had a nominal capital of £50,000; and this was divided into 1,000 shares of £50 each. William Fothergill Cooke of East Tytherly in Hampshire was the leading shareholder with 510 shares; and the other shareholders also resided in England, mainly in or near London. The quarry was worked underground; and a visitor reported in March 1873 that it was dangerous for any stranger to go into the tunnels alone, 'for by for missing a turning one might be precipitated down one of the shafts sunk for ventilation or drainage'. The quarry proved very successful, and its output increased from 397 tons in 1861 to 8,600 tons in 1882. In the latter year this mine was the fifth largest producer in Ffestiniog.[44]

By 1873, slate quarrying was firmly established as a major industry in the Llanberis district. The physical setting, with its lakes, mountains and passes, was highly attractive; but Llanberis itself, which was largely owned by the Ruthin Charity Trust, remained drab. Most of the inhabitants were workmen employed in the slate quarries; and their 'industrial village' was commonly regarded as 'a poor place'. On the eastern side of the valley was Elidir Mountain, which consisted largely of slate rock and extended to Penrhyn Quarry and Bethesda; interest was expressed in the possibility that the two quarries might one day meet.

In 1843 the number of workers at Dinorwic Quarry was 1,900; and by 1873 the number had increased to 2,850. During the period 1832–43 Wellington and Victoria Quarries were developed; they had formerly been known as Hafod Owen and Diphwys. The 'Old Quarry' now came to be known by the name Garrett. The Victoria and Wellington Quarries were near the lake; and the Garrett, Braich, Matilda and Sophie Quarries were above them.

The early tramway across Fachwen to Port Dinorwic became inadequate; and a new line with a gauge of 4ft was built in 1841–2 along the shore of Llyn Padarn. The quarry wagons had to be placed on separate vehicles; and at Port Dinorwic, where the line was several hundred feet above sea-level, the wagons were sent down a steep inclined plane worked by ropes. In 1848 the locomotives *Fire Queen* and *Jenny Lind* began to operate on the new Padarn Railway. Until workmen's trains were introduced in August 1895, the men used to travel along the line on 'velocipedes'. These were trucks for eight persons, and were propelled by means of foot pedals. When Griffith Ellis died in 1860, it was said that in almost a half-century of his superintendence of the quarries the number employed had risen from about 300 to over 2,400. During the last decade of his office, too, Port Dinorwic had been extended by the construction of quays and an outer dock.[45]

Thomas Assheton Smith died in October 1858; and under the will of his widow, George William Duff (1848–1904) succeeded to the estate on his coming of age in May 1869. Duff assumed the additional name of Assheton Smith. The workmen of Dinorwic Quarry congratulated him on his coming of age. They informed him that his great-uncle the 'late Mr. Assheton Smith paid attention and respect to his Quarrymen and used to say of them, "That such people as his, could not anywhere else be found." ' They continued: 'We, the present Quarrymen, are confident that we have not done, nor will do anything that may induce you, Sir, to think less of us, but that you will always find us quiet, honest, and obedient; and we believe that while we continue so, we shall find you a kind, generous and honourable Master.' In 1873 it was possible to write: 'Within the remembrance of persons now living the whole concern was one chaos. At that time it was probably worked profitably, for the time being, but certainly not with an eye to the future.'

In 1870 new workshops were erected at Dinorwic Quarry. They included a foundry, a smithy, a fitting shop, a joiner's establishment and a machinist's shop; and they made up a large rectangular building, with a central courtyard 80yd long and 60yd wide. The buildings faced Vivian Quarry, which was said to have been worked from 1809. All the machinery for Dinorwic Quarry was made on these premises. The work done there included the manufacture of wagons and the repair of locomotives; and the machines were driven by line shafting from an 80hp waterwheel 50ft 5in in diameter. This wheel was installed in 1870; and it continued to operate till 1925, when it was replaced by a Pelton waterwheel. The Dinorwic workshops now form the nucleus of the Dinorwic Quarry Museum, which was opened in May 1972.[46]

On the western side of the Llanberis Vale there were several smaller quarries. Nant Peris had been worked by Lord Uxbridge in the early capitalist period; Glynrhonwy Quarries had been leased to various partnerships by Lord Newborough; and Cefn Du Quarries were worked by sub-lessees of the Cilgwyn Company. These quarries never became large-scale concerns. Carriers' accounts are extant for the Glynrhonwy Slate Quarries between October 1835 and March 1839; and this MS illustrates the conditions governing slate transport from Llanberis to Caernarvon before the opening of the London & North Western Railway branch-line in 1868. Most of the carriers were farmers and cottagers of the neighbouring parishes; and their rates of carriage from Llanberis to Caernarvon in 1835 were as follows:

			Tons	*Cwt*
1st duchesses	15s	per thousand	3	0
2nd duchesses	12s 6d	„ „	2	15
1st countesses	10s	„ „	2	0
2nd countesses	7s 6d	„ „	1	15
1st ladies	6s 3d	„ „	1	10
2nd ladies	5s	„ „	1	5
1st doubles	4s 6d	„ „	1	0
2nd doubles	3s 9d	„ „		15
Moss	7s 6d	„ „		
Rags	5s 0d	per ton		

The places from which the carriers came included Bryngląs, Glynrhonwy, Penybryn, Glanrafon, Cwm y Glo, Talysarn and Penybuarth. The coming of the railway reduced transport costs by about half. Fachwen slate carriers' accounts also exist for short periods. Between 11 July and 24 November 1836, for example, the carriage consisted almost entirely of first, second and third quality rags.[47]

An advertisement inserted in *The Times* and the *Mining Journal* in 1858 described the two Glynrhonwy Slate Quarries, which were held on lease by Captain J. E. Hussey Taylor. The purpose of the advertisement was to obtain 'a partner or partners with capital'. The lease dated from November 1856, was for twenty-one years, and was renewable on payment of £600. No rent was paid; and the royalty of 2s per ton was said to be 'greatly below the average'. The disadvantage of high transport costs, the advertisement claimed, had been overcome by 'an excellent road' intersecting the workings; and the cost of cartage to the shipping quays at Caernarvon was only 4s per ton. The two quarries were named as Chwarel Fownog and Chwarel Fain; and their depths were given as 60ft and 30–40ft respectively.[48]

Cefn Du was one of the most important of the minor Caernarvonshire quarries. In the early nineteenth century it was worked by John Evans and his partners; and in 1812 they sub-leased it to Thomas Bulgin and his partners. From 1878 to 1928 it was one of three quarries worked by the Llanberis Slate Company Ltd. Before the formation of this company the quarry had been owned by a Glasgow merchant named James Wotherspoon. In November 1878 Wotherspoon declared that he had the lease from the Crown, and that he had purchased the quarry from Messrs Stephens in 1874. Wotherspoon claimed that he had spent over £18,000 on it, and that he had in two years increased the royalties from about £100 to £900 per annum; but he still owed the vendor £4,000. He was handicapped by the depression in the slate industry which began in 1876; by the difficulties in opening the east side of the quarry; by lack of capital; and by the shortness of his lease. He therefore sought amalgamation with two smaller quarries in the neighbourhood: these were called the Goodman and the Cambrian, and were worked by T. H. Goodwin Newton of Henley in Arden. Through this amalgamation the Cefn Du Quarry acquired free communication to a siding on the Llanberis branch line of the London & North Western Railway. The lease was renewed for thirty-one years, and the two partners formed the Llanberis Slate Company Ltd. Wotherspoon became secretary, T. H. Goodwin Newton chairman, W. B. Jeffrey manager and John Menzies managing director. The registered office of the company was at Coed y Ddol, Llanberis. Lord Newborough owned the Cambrian Quarry, and the Governors of the Ruthin Charities owned the Goodman Quarry.[49]

The nominal capital of the company was £60,000, divided into 6,000 shares. In February 1879 the company issued 100 debenture shares of £100 each, bearing interest at 6 per cent per annum for seven years. The prospectus described the advantages of the concern. These included the formation of the ground, which sloped upwards from Llyn Padarn. The slope facilitated economical working, there was abundant tipping-space, and the London & North Western Railway ran along the shore. The slate galleries, which had been formed at twelve different levels, were connected with the company's private siding by hoisting apparatus, several miles of tramways, and a system of self-acting and other inclines. In 1882 Cefn Du Quarry employed 197 men; and the output for that year was 5,640 tons. Profits were small. In 1879 the profit was £357 17s; but in 1880 it was only £1 16s 6d, and in 1881 it was £165 10s 9d. In 1883 the company bought a nearby quarry known as Chwarel Fawr; and in 1886 it took over the Bryn Mawr and Bwlch y Groes Quarries.[50]

In the district of Bethesda, Penrhyn Quarry dominated all others; and the only smaller quarries of any significance were Pantdreiniog, Tanybwlch and Bryn Hafod y Wern. A description of Bethesda in 1873 revealed that at that date the village contained one church and nine dissenting chapels. It was said to consist of 'one long street and a mass of straggling houses upon the easterly side of the hillside'. The setting, however, was picturesque; and the beauties of the district included Nant Ffrancon, the River Ogwen, numerous mountain lakes, and such mountains as Carnedd Dafydd and Carnedd Llewelyn.

Pantdreiniog and Tanybwlch Quarries both dated from the early nineteenth century; but Tanybwlch was then abandoned until 1862, when it was acquired by a London company the Port Bangor Slate Company Ltd. The slates were blue mottled in colour, and in 1873 the number of men employed was twenty. Pantdreiniog was taken over by the Bangor and Pantdreiniog Slate Company Ltd in 1873; and in 1882 it employed thirteen men.[51]

The slates from these small quarries had to be carted to Bangor quay or Bangor railway station; and this cost 3s to 3s 6d per ton. Bryn Hafod y Wern Quarry lay on Crown land at the foot of Meol Wnion in the parish of Llanllechid; and it was owned by the Royal Bangor Slate Company Ltd, a London company formed in 1845. Several attempts were made to link the quarry with Aber by means of a branch from the Chester & Holyhead Railway. The first discussions about this plan were held as early as 21 October 1846, when the Chester & Holyhead Railway was still being constructed. The last took place in November 1854, when the matter was indefinitely postponed for lack of funds. In 1852 the company changed its name to the Bangor Royal Slate Company. The slates were blue and green in colour, and the company always hoped that they would find the slate bed which Penrhyn Quarry worked. In 1852 an old quarryman alleged that he had traced the course of this vein through the Crown lands; and in May of that year he was allowed to begin digging. His efforts, however, proved unsuccessful.

Conflicts with Colonel Pennant were frequent. At one time the company thought of reopening the old question of the Pennants' alleged encroachment on Crown land; but in December 1851 they decided on legal advice not to use the evidence of an old quarryman named Parry, 'as such a course might prove prejudicial by its being made to appear that such had been obtained from Parry by intimidation.' Another source of dispute was the supply of water to Bryn Hafod y Wern, as this was obtained from the River Caseg which ran through the Pennant lands. The company also had to discharge their waste water; and on 5 April 1852 it was reported by the manager that Colonel Pennant's tenants were objecting to the increased stream of waste water flowing through their lands. Reference was made on 28 June 1854, to 'the Peculiar, and very dependent, position in which the company stood with Colonel Pennant.' Because of this they judged it advisable, 'in order to secure the future enjoyment of the water from the River Caseg, and its discharge from their works (as at present) to do everything in their power to conciliate that Gentleman—where it could be affected with Consistency, and without much sacrifice to the Interests of the Company.' To avoid a dispute with Colonel Pennant, they agreed to rebuild part of a boundary wall.

In 1873 Bryn Hafod y Wern employed 100 men. The quarry was worked by water-power: there was a waterwheel 30ft in diameter, and a long tunnel brought the water from the River Caseg seven miles away. The water-supply sometimes failed to reach the quarry; and this caused a stoppage of the works. In

1882, when the output was 2,198 tons, the number of men had fallen to sixty-five; and in 1884 the quarry was closed.[52]

In 1826 William Francis became head manager of Penrhyn Quarry in succession to James Greenfield; and by 1835 he was receiving 'for superintending at the Slate Quarries' a monthly salary of £10 16s 8d. Francis was responsible for a vast drainage scheme which carried off water from the quarry's long tunnels and shafts; and John Francis, one of the managers, invented a slate-dressing machine which was worked by treddle. The number of employees in January 1835 was 1,608; and by December of that year it had increased to 1,721. George Hay Dawkins Pennant died in December 1840; and he was succeeded by his son-in-law the Hon Edward Gordon Douglas (1800–86), who was a member of an aristocratic family in Scotland. Douglas immediately assumed the name of Pennant; and in August 1866 he was created Baron Penrhyn. In 1863 the number of men employed at the quarry was approximately 3,000; and in 1869 the output was 93,000 tons.

The appearance of Penrhyn Quarry at that date was vividly described by a contributor to the *Mining Journal*:

> The works are carried on both above and below the level of the original surface. The latter works are first reached. You stand on the brink of an open pit, with irregular sides, jagged with projections of slate rock. Above you and in front rise successive scarpings of the mountain-side. The horizontal lines which, at a distance, appear to stretch like cobwebs at even intervals across the upper workings, are now discovered to be platforms, running from end to end, each forming the base of a separate quarry, along which is a tramway on whose lines horses and trucks are visible. Between these lines, on the vertical face of the rugged rock, men work suspended by ropes slung from the platform above, detaching blocks of slate rock which fall or slide on to the platform below, and are then removed to the wings, where the slate blocks are selected, and the rubble tilted over.

In 1873 the quarry employed about 3,500 men; but in 1882 the number had fallen to 2,809. The output for the latter year was 111,166 tons; this compares with an output of 87,429 tons at the same date in Dinorwic.[53]

7
THE MECHANISATION OF THE QUARRIES

One essential for the development of a quarry was an incline to bring the slate down to where a tramway could start. According to D. C. Davies, who was writing in 1878, the best kind was one with a double line of rails from top to bottom. The gauge was usually 1ft 11½in; and this was also the gauge of the Festiniog Railway, the Croesor Tramway, the Gorseddau Junction Railway and the North Wales Narrow Gauge Railway. This narrow gauge was very convenient for tracks running through the quarry, the sheds and the yards.[1]

In August 1864 Allen Searell sent Edward L. Goetz plans, sections and an estimate for the proposed redevelopment of the Royal South Snowdon Slate and Slab Quarries in the parish of Beddgelert. The total cost was estimated at £30,000; and £1,000 of this was allocated for making inclines from the quarry to the railway.[2] A list of the plant held about 1880 by the Moelwyn Slate Quarry near Tanygrisiau included the following items:

A Tramway laid on Inclines and Levels about 2,100yd
Double on Incline and first four levels, the rest single.
4 Drums 6ft diameter by 11ft 9in
2 Drums 5ft diameter by 11ft 9in fitted with wire ropes, chains, hooks, shackles and about 250 iron rollers.
Machine sheds fitted with 6 saw tables and 7 dressing machines
Powder Magazine

1 Water-Wheel 40ft diameter, 4ft breast, with conduit to same from Lake
Carpenter's shop fitted up
Blacksmith's shop fitted up
1 Patent Saw-Sharpener
4 Crowbars
18 Rock Drills
1 Grindstone
2 Wheelbarrows
6 Lamps
2 Weighing Machines

The tramway from the quarry joined the Festiniog Railway near Tanygrisiau;[3] and this list can be compared with a list of the plant belonging to Cilgwyn Quarry in July 1827. This included:

2 water-wheels
4 whimseys
7 Cars and their wheels
2 weights
3 Turntables [4]

Advertisements for the sale of quarries usually provided information about the quarry machinery to be sold. The sale of Croesor Quarry was announced on 13 May 1848 in a notice headed 'Important to Capitalists'. It was confidently asserted that there was 'the greatest facility for carrying on operations at all the Works', and that this could be done 'with little expense, as few or no Machinery will be required.'[5] When Croesor Quarry was being offered for sale again in 1874, it was said that a great deal of capital had been spent on the works. Besides opening up new chambers, the owners had introduced 'very extensive Saw mills, 2 powerful Water-Wheels, Sawing Tables, Planing Tables, Slate Dressing Machines, Long Tunnel for drawing and working the quarry, extensive railway inclines, waggons, 2 12 h.p. locomotive steam engines . . . workshops and stables.'[6]

The need for water-power was recognised in most quarry

advertisements. The Rhiw Quarry, in the parish of Penmachno, was up for sale in October 1828; and one of its attractions was the 'excellent fall of water' near the quarry, which would 'turn Mills and other machinery, for making mantelpieces, gravestones and writing-slates etc'.[7] In order to store the water when the machinery was not in use, it was generally necessary to build reservoirs. The largest in Merionethshire were Barlwyd, Bowydd, Ffridd and Conglog lakes. Water was conveyed to the mills along wooden troughs supported by stone pillars. The use of machinery resulted in the building of machine houses or mills. These were first introduced in the Ffestiniog area, where they replaced the small sheds in the early 1850s; and they enabled quarries to coordinate their manufacturing processes more effectively.[8]

In 1878 D. C. Davies described a typical machine-house as a building with 6 saw tables, 2 planing tables, 16 slate-dressing machines and 4 turn-tables. All these machines would be worked by a waterwheel 30ft in diameter; and the total cost inclusive of shafting, straps and gearing would be about £1,978.[9]

About 1807 Robert Meyrick Humphreys erected a steam pump at Hafodlas Quarry. Articles of co-partnership which were drawn up in 1812 declared that he had 'already at his own expense erected upon the said Quarry a Steam mill for the purpose of clearing the said Quarry from Water and . . . thereby agreed to lay out at his own like expense a further Sum of Money not exceeding together with the said Steam Mill sum of £1,000 in erecting Windmills or any other necessary and proper Engines, Machines or Erections proper and necessary for clearing the said Quarry from the water now contained therein and keeping the same free and clear wherefrom during the continuance of this agreement.'[10] A 'wind engine' or windmill was erected at Braichyrhydd Quarry, one of the quarries in Nantlle which John Evans and his partners were trying to develop in 1827. John Evans wrote to John Price on 17 April 1827:

> I am surprised that any one should be so ignorant as to inform you that the wind engine will be of no use, for we cannot work the Quarry without it and as to its being blown

Page 155 (*above*) The incline from the terminus of the main line at Port Dinorwic; (*below*) Rhiwbach Quarry incline, showing an accident on the line, about 1900

Page 156 (*above*) The Dinorwic quarrymen's train, the *Amalthea*, about 1900; (*below*) Port Dinorwic, showing a steamer in dry dock

> down I will engage that it will stand for ten years. You are so dreadfully imposed upon.

In June 1806 John Evans and his partners had planned to erect a windmill at Cilgwyn Quarry: it was to be built by a blacksmith for £120, and was to raise water from the bottom of the quarry.[11]

Steam-power gradually replaced water-power; but in the Ffestiniog mines it was not used until the 1850s. In 1856 a steam engine was being used to raise rubbish in the Middle or Rhiwbryfdir Quarry. In the 1860s steam-power came into more general use in the Ffestiniog district; and by 1873 all the machinery at Diphwys Casson Quarry was being worked by steam. Steam-engines were faster than water balances; and so they were often preferred for lifting rubbish from the lower reaches.[12]

In 1873 Llechwedd Quarry had 'complicated workings', with large and small waterwheels on all sides but no steam-power. The inclined shaft communicating with the underground workings had three tramroads laid on it; and each of these connected with a flooring. At the top of the incline were three drums with coils of steel wire attached to them. These were used to hoist the rock from the levels; and the power was supplied by a 'gigantic water-wheel'.[13]

In Bryneglwys Quarry slates were brought from different levels by means of two large waterwheels; and in 1873 there were some inclined chains for hoisting loaded wagons from the pits. The wagons were attached to a wheel which traversed the inclined chain. Another labour-saving device at this quarry was the system of tramroads: these were connected by turn-tables so that the wagons could be forwarded to any part of the bank with 'little labour and much expedition'.[14]

In Nantlle the quarries were badly off for water-power. In Dorothea Quarry a large amount of capital had been expended by 1864 on plant and machinery. The quarry at that date possessed the following plant:

> 3 steam engines, nearly new, with drums, pulleys, chains,

> and all connecting gear for the purpose of hauling up the produce, 3 large water-wheels for pumping out the water, and working sawing and planing machines, 2 sawing and 1 planing machines, tramways, trucks and wagons, offices, weighing machines, dressing sheds, smiths' and carpenters' shops, and various other buildings and cottages erected for the accommodation of the workmen.[15]

In 1879 the Llanberis Slate Company, which was the company running the Cefn Du, Cambrian and Goodman Quarries, had the following machinery:

> 4 fixed steam engines, one tramway locomotive, steam rock drill, 2 water-wheels and a turbine hoisting apparatus, drums, steel and iron wire ropes, and a large number of slate and rubble waggons, trollies etc.

The buildings included 'substantial and roomy workshops, ranges of slate dressing sheds, weighbridges, engine houses, offices, powder magazines and store houses'. In October 1888 the company installed sawing machinery, which enabled them to produce a large proportion of 'Best quality' slates.[16] Slates and waste rock from underground workings had been hauled up in the early days of the slate industry by wheelbarrows; and later horses had been used to drive whimseys. Horses were still used in some concerns as late as the 1960s.[17]

Machines for dressing slates were invented separately by Nathaniel Matthews, John Greaves and Samuel Holland. The machine invented by Greaves was being used by several of the largest quarries in 1873; and it is still being used in 1972.[18] In 1873, however, slates in Nantlle and indeed in the whole of Caernarvonshire were being worked not by machinery but by hand.[19] Rhiwbryfdir Company had a steam engine in 1856 to drive Matthews's slate cutting machines.[20]

Most of the underground workings were lit by candles. In October 1896 some of the slate proprietors at their fourth annual congress visited the Oakeley Quarries, which at that time included Holland's Quarry, Palmerston Quarry and Rhiwbryfdir Quarry. Armed with lanterns they descended by the water balance shaft and entered the K floor about 1,500ft down: 'In single file the party was taken something like one and a half miles, along this subterranean passage, here and there passing through chambers, which were being worked. Here and there, above and below, could be seen quarrymen, with the aid of a candle as their only light, carrying on their daily avocations.'[21] In 1937 the underground inclines, landing and most of the tunnels were illuminated by electricity; but the bulk of the workings were still lit by 'primitive tallow-dip'. In 1949, an efficient type of electric battery cap lamp was introduced into the larger Ffestiniog concerns.[22]

In 1896 Llechwedd Quarry was declared to be 'in the way of machinery about the most up-to-date in the world'. The manager at that time was Warren Roberts, who was an engineer; and the first application of electricity to slate quarrying was made in Llechwedd Quarry in 1890. Quarry proprietors visiting this quarry in 1896 noted especially the electric pumps in the mine and the dynamo house, which contained two dynamos of 60 and 120 amps, with power from a Vortex turbine of 28hp. By 1929 this was developed into a complete hydro-electric installation. Water-power was supplied from a reservoir above a power station in which there were two turbines. From the power station the current was distributed to air-compressors, winding gear, pumping plant, mills and workshops.

Electric drilling plant was used in quarrying the slate; and electrically-driven pumps on various floors of the workings helped to drain the mine, which had drainage levels constructed to run all superfluous water into sumps, from which it could be pumped either to a higher sump or to the surface. Outside the mine, the winding gear for carrying slate and rubbish was also operated by electricity. The light railway system in the mine was electrified; but the main haulage work on the surface levels was carried out by overhead trolley locomotives, one of these being a steam locomotive converted to electric power.[23]

In 1896 Dorothea Quarry obtained and accepted an estimate

of £6,700 for electrical plant for working the mill and pumps.[24] The urgent problem at Dorothea was drainage, since the pits were set on the margin of a lake and went down to a depth of 400 to 500ft below lake-level. A Holman Cornish beam engine with a 4in bore was installed in 1904; this was capable of pumping 600 gallons of water a minute from a depth of 500ft, so that running continuously it lifted almost 1,500,000 gallons in 24 hours. In 1955 the engine was replaced by a 60hp Beresford 8in centrifugal pump; but it remained in situ as a memorial, being cared for by the Industrial Steam Preservation Group.[25]

The use of electricity in the slate concerns largely revolutionised the industry. Until the beginning of the twentieth century many quarries had been almost entirely dependent on water-supply, which meant that any interference with this caused a stoppage of the works. In May 1802 quarrymen working Hafodlas Quarry wrote to the Rev R. M. Humphreys:

> This is to inform you that we are again deprived of the Stream of Water from Coedmadog and Consequently the Pumps and Engine cannot work at Hafodlas Quarry which fills up with water very rapidly and has put an End to the working of the said Quarry which is a very great loss.[26]

No doubt this was one of the main reasons why Hafodlas Quarry was one of the first to install a steam engine. Bryn Hafod y Wern Quarry was also in a weak position, as its water-power came from a small lake, called Ffynnon Gaseg, which belonged to Lord Penrhyn. A water-course led from this lake to two reservoirs; and from the lower reservoir pipes and troughs carried the supply necessary to operate the lifts, the water balance, and the pumps which lifted water from the quarry. Unfortunately for the proprietors, Lord Penrhyn was able to cut off this supply at any time. In May 1889 the solicitors of the Bangor Royal Slate Company were informed that 'under Lord Penrhyn's directions Col West, as agent of the Penrhyn Estate, stopped the water flowing through the Quarry over Lord Penrhyn's enclosed lands on and from Monday 27th inst.' This stoppage was 'intended to be permanent'; and it marked the end of the quarry's working life.[27] Dry spells could exhaust water-supplies, and stoppages could last several weeks. In severe winters reservoirs and troughs were often frozen, so that waterwheels could not function. Water-power was later harnessed to drive machinery for sawing-tables and dressing-machines; but at best it was irregular and could only be transmitted over short distances.

The development of the slate industry was speeded up by the use of steam-power; and steam superseded water for most quarry operations, though water balances remained in use. With steam came the use of compressed air, and this was used for operating winches and drills. Most of the compressors were worked by steam; but a few, including that at Llechwedd, were powered by water. In 1896, when the proprietors visited Llechwedd Quarry, they saw an air-compressor which was driven by an 8in canvas belt from a turbine under the same roof. The compressor ran at a speed of 90 to 95 rpm, and was capable of providing air at 60lb pressure per sq in to work two 3in rock-drills. The belt ran at a speed of about 24mph and the turbine was one of 35hp using 320cu ft of water per min. The spindle of the turbine ran at 575 rpm, and supplied motive power to two mills. By 1902 compressors worked by steam were in use at Oakeley, Craig Ddu and Rhiwbach Quarries. These rock-drills were used chiefly on the top-rock; when they were used in underground chambers, they were called channellers.[28]

The drawback of steam was the difficulty of transmitting the power to distant parts of the quarries; this problem was solved only by the introduction of electricity. After Llechwedd Quarry had installed electricity in their works, other quarries followed their example. In 1900 the Votty and Bowydd Quarries obtained an electricity supply from the Yale Electric Power Company Ltd, and used it for pumping and hauling slates up inclined planes. In 1913 these quarries made an agreement with the Yale Electric Power Company Ltd, whereby they paid an annual minimum sum of £1,600 for the supply of power.[29]

Moses Kellow (1862–1943), the manager of Croesor Quarry who arranged the amalgamation of Park and Croesor Quarries in

1895, described in his autobiography how in 1901 he prepared a scheme for the electrification of Croesor Quarries. He proposed to obtain power from Cwn Foel Valley, which was the source of Croesor River; and in defiance of British electrical manufacturers he decided to use plant with a 'three-phase alternating current' giving a rotary motion, which was made for him by Kollen & Company of Prague. Kellow, who was regarded as a 'fearless innovator' in the engineering world, introduced a 30hp electric mining locomotive for haulage in the main tunnel: this was the first electric locomotive to be used in Wales. Kellow also introduced reversible winches, each of which was driven by a 10hp motor. These winches were geared to give a three-ton pull on a rope for loading blocks; and they had an alternative high speed for driving the blocks along the floor. The shunting-tracks, sidings, inclines and tunnels were lit by electricity; and each of 2,000 had two arc lamps, each of 2,000 candle power, suspended overhead. Kellow also invented an improved type of planing machine which enabled the slate to be planed from six to ten times faster than before.

Drainage was achieved by Kellow through a series of electrically-driven centrifugal pumps which raised 400 gallons of water a minute; Kellow replaced the hand-drills by a Kellow drill which used high water-pressure and was 'absolutely dustless'. This reduced the time required to make a 7½ft hole from 10 hours to 1½min. Some quarries, including the Cambrian Quarry at Llanberis, used a rock-drill known as the Burleigh Rock-Drill; this bored a hole 2in in diameter and 10ft deep at the rate of 1ft in seven minutes. An earlier drilling machine was Dixon's boring machine. When this machine had drilled a hole 3in in diameter to a good depth, a charge of from 15 to 20lb of powder was inserted; and the subsequent explosion brought down a large amount of rock, broken into large blocks.[30]

In 1905 the North Wales Power & Traction Company began to supply electricity to mines and quarries at rates of from ⅞d to 1½d per unit. One of the transmission lines went across the Lledr Valley to Ffestiniog to supply Oakeley Quarry, another went via Penypass to Llanberis to supply Dinorwic Quarry, and a third went via Drwsycoed Pass to Nantlle Vale to supply Penyrorsedd and other quarries in the Nantlle area.[31] In September 1906 electric plant was installed at Oakeley. The power, which was generated at the Cwm Dyld Power Station of the North Wales Power Company, totalled 1,930hp; and it was utilised for driving hauling motors. This was the first large-scale attempt to use electrical power on quarry inclines; and the underground inclines were also converted to electricity.[32]

Rhiwbach Quarry, although a comparatively small concern, was investigating new techniques in 1906: the pamphlets consulted included *Electric Rock-Drilling Machinery* by A. L. Stevenson (Cleveland 1893) and *La Puleggia Penetrante* by A. Monticolo (Rome 1899). In correspondence with firms in Belgium and France, the manager inquired about wire-saws; but experiments with these saws proved unsuccessful.[33]

The open quarries were slower than the mines in introducing modern quarrying methods. Before 1912 drilling was carried out by hand at Penrhyn Quarry; but in that year pneumatic drills were introduced, and electric power was supplied to the quarry by the North Wales Power Company.[34] In 1912 also motor and haulage gear were attached to many of the old gravity inclines in order to concentrate the waste rock in two or three places instead of in twenty-three tips.

A sawing machine known as Hunter's, which had circular teeth, was used in some of the Llanberis and Ffestiniog quarries in the 1870s. About 1850 J. W. Greaves had invented another sawing machine, which combined a movable truck and a circular saw; as the truck moved forward the saw cut through the blocks. The Llanberis Slate Company used a steam engine in the 1920s to drive their saws. The saw shed was 116yd × 14yd and 10ft in height; and the engine house was inside the shed. Steam power was also used to drive the saws at Dorothea Quarry.[35] At Aberllefenni Quarry near Corris experiments were made with a slate-splitting machine invented by a man named Hughes; but elsewhere the process of splitting was still carried out by hand.[36]

Diesel and petrol locomotives were introduced into the Ffestiniog mines in the 1930s; and in March 1932 an aerial cableway was built over the eastern section of Llechwedd Quarry. This was a further advance in mechanisation, as the aerial way carried waste rock at the rate of 1 ton per minute to a distance of 1,388ft from the place where it was worked. Having been lifted from a depth of 100 to 200ft, a 4 ton load could be transported and tipped. This helped to overcome the shortage of tipping ground in the immediate vicinity.[37]

The introduction of gunpowder increased the hazards which the workers faced; but the use of the safety fuse lessened the risk, as this could be ignited at a safe distance from the charge. Dynamite, nitroglycerine, gelignite and other powerful explosives were used to heave and displace the rock without shattering it into fragments. Gunpowder was used for blasting the slate, and more powerful explosives for driving tunnels and blasting igneous rocks. The holes into which the charges were inserted were at first made by hand-drills; but later compressed air-drills were used. Rules for blasting were introduced in most quarries: these insured that it was carried on throughout the quarry or mine at certain prearranged times, so that the quarrymen could shelter in their soundly-built sheds. The men did not return to work until an agreed number of minutes after the explosions; and each group was responsible for making sure that the charges had gone off.

In 1895 Dorothea Quarry published its printed rules for blasting in English and Welsh and submitted them to C. Le Neve Foster, who was responsible for the inspection of quarries. Foster amended the rules, so that they appeared in the following form:

1 Blasting is allowed hourly.

2 Signals: a red flag will be hoisted upon a conspicuous spot for seven minutes during the blasting operations. Also, a steam whistle will be blown thus:

ONE whistle means that the men must retire to a place of safety; after the lapse of two minutes, two whistles in rapid succession mean that the fuse must be lighted and after the lapse of five minutes

THREE whistles in rapid succession and the lowering of the red flag mean that all Blasting has ceased and that the men can return to their working places.

3 When Blasting takes place during the Dinner hour or immediately after working hours are over, the above signals (Flags and Whistles) must be strictly observed. Five minutes, however, must be allowed from the time the whistle to cease work has been blown before the first of the above signals is given.

4 When the Whistle to cease work is blown before 5.30 p.m. no blasting can take place after 4 p.m.[38]

One of the most notable of a series of 'big blasts' which took place in several quarries at the end of the nineteenth century was the removal of a large pillar known as 'Talcen Mawr' at Penrhyn Quarry. The event took place on 27 April 1895; and it was followed by similar huge blasts at Cilgwyn Quarry and Dinorwic Quarry. In January 1896 at Cilgwyn Quarry a 'majestic dyke' which was about 150yd high and about 120ft in diameter was demolished by explosives from the Nobel factory in Glasgow. The rock had hindered the development of the quarry; and over 100,000 tons of granite were removed by force of the explosion. At Dinorwic Quarry in 1896 explosives were used to remove a ridge of hard granite known as the 'Big Cock'; and the event was watched by the Assheton Smiths, Lord Penrhyn, Lord Mostyn, and E. A. Young and hundreds of quarrymen.[39]

In 1902 the Prince and Princess of Wales visited Bangor; they were entertained by a big blast at Dinorwic Quarry, and were the guests of the Assheton Smiths at Vaynol House. The Princess was presented with a slate casket containing a key; and when she used this key a whistle warned the workmen to take shelter. Over two tons of black powder were used in the explosion; and an eye-witness account described how 'the light puffs of smoke from the fuses . . . burned their way to the monster charge of explosive, and then, heralded by a vast plume of smoke the front of the hill opened and changed its appearance as if by magic. The roar of the explosion shook the mountain, and, as the black clouds from the charge dispersed slowly skyward, volumes of dust ascended and suggested still another burst, while tons and tons of slates rolled and rattled down the terraces of the quarry.'[40]

Technical improvements in the slate industry increased output; but the use of machinery brought new problems, such as an increased incidence of silicosis, and a need for more complex safety regulations. Before 1894 the safety legislation governing slate mines and quarries was very unsatisfactory. The Metalliferous Mines Regulation Acts of 1872 and 1875 fixed at 12 years the age below which boys were not to be employed underground; and in 1900 this was raised to 13 years. Mines and quarries, however, were under a statutory obligation to provide for the safety and health of the workers only in so far as the Factory Acts applied to them. These affected only the larger quarries; and their effect was limited to such matters as the fencing of machinery and the inspection of slate-dressing sheds. In 1887 the Quarry (Fencing) Act (50 and 51 Vict cap 19) was passed; but in many districts it was not properly enforced. There were no regulations governing the chief causes of accidents at quarries, such as falls of ground and blasting operations. The Quarries Regulation Act of 1889 applied to all slate quarries where more than twenty persons were employed. One of its clauses concerned the notification of accidents; and another concerned the fencing of abandoned quarries.

The Quarries Act of August 1894 affected all excavations more than 20ft deep; and it authorised the responsible minister to make the rules for the prevention of dangerous accidents and to provide for the safety and discipline of the workers. Under this Act the inspection of quarries was assigned to HM Inspectors of Mines; but the Factory and Workshop Act of 1901 transferred this responsibility to the Factory Inspectors. A code of special rules for Merionethshire mines was drawn up in 1900; and two Welsh-speaking inspectors were appointed for the Ffestiniog area.

Two Acts which established the right of the workers to recover compensation from their employers for injuries received in the course of their employment were the Workmen's Compensation Act of 1880 and the Workmen's Compensation Act of 1911. Slate quarrying is a dangerous occupation; and fathers were often reluctant to let their sons follow in their footsteps. The death rates per 1,000 employees for the North Wales slate quarries from 1913 to 1920 were as follows:

	Inside	*Outside*	*Total*
1913	2·90	—	1·17
1914	1·66	—	0·66
1915	3·96	—	1·53
1916	4·94	—	1·80
1917	4·38	1·08	2·17
1918	2·43	—	0·81
1919	0·56	0·35	0·43
1920	1·75	0·29	0·87

In the years 1901–11 there were fifty-four deaths in the North Wales slate mines and ninety-eight deaths in the slate quarries. The numbers of persons injured through accidents in the years 1908–12 were as follows:

Quarries

Year	*Persons Injured*	*Number Employed*	*Rate per 1,000 Employed*
1908	903	8,056	112
1909	970	7,813	124
1910	813	7,867	103
1911	701	7,732	91
1912	622	7,194	86

Mines

Year	*Persons Injured*	*Number Employed*	*Rate per 1,000 Employed*
1908	174	3,479	50
1909	211	3,327	63
1910	255	3,465	74
1911	309	3,697	84
1912	246	3,417	72 [41]

Falls from the working faces of the quarries of great depth were difficult to avoid entirely, because so much depended on personal factors; but certain precautions, such as the secure anchorage of ropes, lessened the risk. During the nineteenth century the accident rate in the slate industry was comparatively high. There was little state regulation to control conditions; and the workers themselves sometimes ignored such rules as those regulating blasting. Because of a lack of stretchers, injured men were often carried down roughly; and hospital treatment was often inadequate. During the twentieth century, however, the fatality rate has declined.[42]

In 1964 the engineering firm of McAlpine and Company Ltd acquired a controlling interest at Penrhyn Quarry; and a policy of modernisation was initiated. The programme began with the construction of an access road from the main splitting area to the highest galleries and to the bottom gallery. The difference in level between the top and bottom galleries was 1,200ft, and the road connecting these sections was about 17,500ft long. The cost of the roadway was approximately £100,000. The development of this road enabled mechanical equipment to reach the two parts of the quarry where the major reserves lie. Work, in 1972, is nearly completed on the modernisation of the Fullersite Plant, which produces 300–400 tons a week of very fine inert slate powders for the plastics industry and others.[43]

In 1964 new developments were carried out at Dorothea Slate Quarry; and £25,000 was invested in modern equipment, chiefly to remove a large quantity of overburden covering the slate rock. One innovation was the introduction of bulk-blasting: vertical shot-holes 4in in diameter were drilled with Halco-Fenwick machinery; and the waste rock was removed by an Atlas Copco L M 30c rocker shovel, which loaded the waste into a Thwaite Sprite dumper. Suitable pieces of rock were hoisted to the surface in narrow-gauge trucks by blondin aerial ropeways manufactured by Henderson's of Aberdeen. Two blondins were used in the quarry. The men were not allowed to ride on the rope-ways, but had to use 200ft ladders. The slates from the blondins were taken to the weighbridge and then to the slate mill along a narrow-gauge line, the wagons being pulled by a horse. In 1964 the North Wales slate industry was said to be entering 'a new era of development'; and the owner of Dorothea Quarry declared that 'the day of the man with a pick and shovel' was over. In 1969, however, both Dorothea Quarry and Dinorwic Quarry ceased operations.[44]

8
IMPROVED TRANSPORT AND ITS CONSEQUENCES

The distance of most slate concerns from the sea, and the mountainous nature of their situations, produced serious obstacles to the early development of the industry. Like coal, slate was a heavy and bulky substance to transport; and unlike coal it was of little use if it became damaged. The transporting of slates in carts was an improvement over carrying them in panniers on mules and horses; and in the Bethesda area large wagons could carry as much as 2½ tons at a time.[1] Penrhyn Quarry was about seven miles from the coast; and the Llanberis quarries were about eight miles from Caernarvon and Port Dinorwic. The Nantlle quarries were about ten miles from Caernarvon; and the Ffestiniog mines and quarries were about ten miles from Portmadoc. The slate districts of Denbighshire and Montgomeryshire usually sent their output by canal to English towns.

The building of roads was usually undertaken by quarry owners, such as Samuel Holland and Lord Penrhyn. Samuel Holland planned a road to Rhiwbryfdir Quarry in 1821; but it was opposed by Turner and Casson, through whose land it would have run. This opposition was overcome; and when the quarry was sold to the Welsh Slate Company the road was almost complete. Turner and the Casson brothers made a new road from Congl y Wal to Talybont; Holland used this for a while, paying them a toll.[2] The proprietors of Manod Quarry constructed a road from their quarry to Maentwrog 'at a considerable Expense'.[3]

Minor roads were usually built by parochial efforts, although these were often assisted by private enterprise. Major roads, however, were usually developed and managed by turnpike trusts. In 1752 such a trust completed a new road from Shrewsbury to Wrexham; and in 1759 another trust was established to manage the road from Tal y Cafn ferry to Conway. It was not until 1777 that an Act was passed enabling the main roads in Merionethshire to be converted into turnpike roads, thereby improving communications with Caernarvonshire. By 1800 the turnpike trusts were responsible for 1,000 miles of road in North Wales. In 1802 Bangor was linked to Capel Curig and Pentre Voelas; and roads existed from Bala west to Maentwrog, north-east to Corwen and Llangollen; south-east to Llangynog, south to Dinas Mawddwy and south-west to Dolgellau and Barmouth. In 1823 the Holyhead Road through Anglesey was completed, having been constructed under the direction of the Parliamentary Commissioners; this superseded the former turnpike.

Most of the important turnpike roads had been made by 1810; and by providing easier access to markets they were invaluable to the slate industry. The development of coaching, which began in North Wales in the last quarter of the eighteenth century, enabled quarry agents and proprietors to communicate with each other and their customers more speedily. Samuel Holland, for example, occasionally took journeys to 'see and become acquainted with slate merchants'; and he made contacts in Newport, Swansea, Cardiff, Bristol, Gloucester, Plymouth, Exeter, Southampton and London. About 1776 the landlord of the *White Lion* in Chester began to convey passengers every weekday to Holyhead by a 'flying post chaise' for two guineas. From 1808 the mail coach from London to Holyhead ran nightly through Shrewsbury, Oswestry, Corwen and Pentre Voelas.

In 1840 the Royal Commission on the State of the Roads praised almost all the turnpike roads of North Wales. At that date the only railways were horse tramways, but they had taken the heavy slate traffic of the quarrying districts off the roads. Most of the turnpike trusts expired in the 1860s and the 1870s. The opening of the Chester to Holyhead railway in 1849 was a very significant event, as it speeded up transport considerably. The mail coach was then discontinued, and most of the coaching services came to an end.[4]

Lord Penrhyn's railway, whose origins have already been described, was completed in 1801. In 1840 steam locomotives were introduced on this track; and in 1876 a new narrow-gauge railway replaced the older one. The gauge of the new line seems to have been 1ft 10¾in, and, except for about half a mile, near the second incline of the old track, it followed a new route, more to the west, running through Tregarth and along the River Cegin valley.[5] Until 1852, because of the distance from Port Penrhyn to Bangor Station, Penrhyn Quarry found it difficult to send slates by railway. In February 1852, however, a branch railway of one and a half miles was opened to link Port Penrhyn with the Chester & Holyhead. The Penrhyn Railway continued to operate a passenger service for the quarrymen until 1951; and in 1962 it ceased work altogether. The line was taken up, and the rails were sold to the Festiniog Railway.[6] Thomas Assheton Smith's tramway, which had a gauge of 1ft 10¾in, was completed in 1824. The line was seven miles long; and it had two inclined planes, one about a mile from Dinorwic Quarry and one near Port Dinorwic. The Padarn Railway was begun in 1841–2; it had a 4ft gauge, and terminated at the top of an inclined plane behind Port Dinorwic. Two narrow-gauge tracks were laid on this inclined plane; and each 4ft wagon contained two sets of 1ft 10¾in gauge rails mounted on a frame-bottom. Each big wagon could thus carry four of the smaller gauge trucks; and these small trucks carried the finished slate. Steam locomotion was introduced in 1848. The first two locomotives, *Fire Queen* and *Jenny Lind*, were built at Northfleet in Kent by A. Horlock & Company. The chimney of *Fire Queen* extended to 11ft 6in above rail level.[7]

Port Dinorwic was connected by a branch of the Caernarvon & Bangor Railway with Chester and the towns of the Midlands. In 1869 the Caernarvon & Llanberis Railway was opened; this was of advantage to the Glynrhonwy Quarries, which had continued to use carriers until this date.

The Festiniog Railway, the most famous of all Welsh narrow-gauge railways, was constructed by James Spooner with the assistance of his 15 year-old son Charles Easton Spooner. Many engineering difficulties had to be overcome, as the line ran high above the valley of the River Dwyryd and crossed the deep ravines on high stone embankments. There were also two tunnels, one of which was 730yd long. Before steam locomotives were introduced, the slate trains were drawn by horses when returning empty up to Blaenau Ffestiniog; but on the down journey they went by gravity, the horses being carried in 'dandy wagons'. On the embankment, horses pulled the trains in both directions.

The Festiniog Railway was built as a single-line track; and it remained one, despite the fact that double lines were authorised in the Act of 1869. It was laid at first on stone sleepers with rails weighing 16lb per yd. In 1851–2, as a result of heavier traffic, the line was relaid with 30lb rails in lengths of 18–21ft; these were laid on wooden sleepers placed in 10lb chairs. In 1869 these rails were replaced by 48½lb bull-head rails 24ft long, which were laid on larch sleepers spaced 3ft apart. This relaying was necessary because the 30lb rails were not heavy enough for locomotive traction. In June 1863 locomotive traction was introduced on the railway, which thus became the first narrow-gauge steam railway in the world.

The locomotives used were two 0–4–0 side tank engines built by George England & Company of Hatcham Iron Works in London; they were called the *Princess* and the *Prince*. The same company later supplied double-bogie articulated locomotives of improved performance; and the first of these, the *Little Wonder* was introduced in September 1869. The line was officially opened for passenger traffic on 6 January 1865; and the stations in the first time-table were Portmadoc, Penrhyndeudraeth, Hafod-y-Llyn and Dinas. In February 1866 a new station was opened at Tanygrisiau, and in 1870 the passenger station at Dinas was closed. The Hafod-y-Llyn station was replaced in July 1872 by one at Tanybwlch.

The first passenger carriages on the Festiniog Railway were all four-wheelers; they included not only closed and open carriages but also one with an 'umbrella roof'. Bogie carriages came into use in 1878. One type had seven compartments, divided into first, second and third class.[8] Criticism of the service provided for the quarrymen by the Festiniog Railway was voiced during an 1895 inquiry into the slate mines of Merionethshire. It was said that the only light was candle-light. More general criticisms were levelled against the narrow-gauge railways; and it was said that there was no sufficient bench accommodation at their stations.[9]

The tonnages of slates carried by the Festiniog Railway from the Ffestiniog quarries from 1848 to 1937 were as follows:

Total for the Years	*Tonnage of Slates*
1848–1852	189,637
1853–1857	236,077
1858–1862	256,467
1863–1867	378,691
1868–1872	594,328
1873–1877	627,545
1878–1882	601,216
1883–1887	529,029
1888–1892	509,110
1893–1897	636,849
1898–1902	619,758
1903–1907	534,406
1908–1912	426,431
1913–1917	293,284
1918–1922	242,688
1923–1927	302,846
1928–1932	290,173
1933–1937	222,247 [10]

The Festiniog Railway divided into two branches at the approach to Blaenau Ffestiniog. The main west line ran to a terminal at Dinas, and the branch or east line ran to Duffws. There were two branches at Duffws: one was used by Maenofferen Quarry, Bwlch y Slaters Quarry and Rhiwbach Slate Quarry; and the other went by an incline to Diphwys Casson Quarry and to Votty and Bowydd Quarry.[11]

Many of the slate companies took advantage of the power given to owners of land by the Festiniog Railway Act to build branch lines to connect their works with the railway. Some companies expended large sums on inclines; and in the case of the Wrysgan incline, constructed in 1844–5, this expenditure helped to cause the decline of the company.[12] Plans were drawn up in 1853 for a branch line through the Machno Valley to the Penmachno quarries; but they were never carried into effect.[13]

The Welsh Highland Railway ran from Dinas Junction, about three miles south of Caernarvon, to Portmadoc. It had its origins in the Croesor Tramway, which served the Croesor and Rhosydd Slate Companies, and the Gorseddau Junction & Portmadoc Railway, which was built by the ill-fated Gorseddau Slate Quarry Company. In 1865 the Croesor & Portmadoc Railway Company was incorporated; and this converted the lower section of the Croesor line into a public railway. The remaining part of the original line remained private property, and was known as the Croesor Tramway. This tramway had three double-track gravity inclines: Blaen y cwm, 835ft long; Upper Park, 1,225ft long; and Lower Park, 1,725ft long. It served several slate quarries, and carried general traffic to and from the village of Croesor. It was auctioned on 14 October 1936, along with the Park and Croesor estates. The lower part of the Croesor Tramway, which with the approach of the Cambrian Railways had become the public Croesor & Portmadoc Railway, was the first section of the later Welsh Highland Railway. The North Wales Narrow Gauge Railway Company was responsible for the scheme to include all the route finally claimed by the Welsh Highland. The North Wales Narrow Gauge Railway Company was incorporated by Act on 6 August 1872; and on 21 May 1877 the portions of the line from Dinas to Bryngwyn and from Tryfan Junction to Quellyn were opened for goods traffic. Passenger traffic over this section began on 15 August 1877. An extension of ¾ mile from Quellyn to Snowdon Ranger was opened on 1 June 1878; and an extension of 2 miles from Snowdon Ranger to Snowdon (Rhyd-Ddu) was opened on 14 May 1881.

The Act of 21 June 1879 changed the name of the Croesor &

Portmadoc Railway Company to the Portmadoc, Croesor & Beddgelert Tram Railway Company. Attempts to link the North Wales Narrow Gauge Railway with the Croesor system failed; the former was operated by steam, and the latter was horse-drawn. In 1921, at a public inquiry by the Light Railway Commissioners, it was said that only freight was carried. On 30 March 1922 the Welsh Highland Railway Company was incorporated, and this company acquired both the North Wales Narrow Gauge Railway Company and the Portmadoc, Beddgelert & South Snowdon Railway Company. The section from Dinas Junction to South Snowdon was reopened for passenger traffic on 31 July 1922; but passenger traffic never ran on the Bryngwyn branch. A new section of 8¾ miles between South Snowdon and Croesor Junction was completed in June 1923.

The railway was never a commercial success: the slate industry in its area declined, and the infrequent train service was unable to compete with road services for tourist traffic. Passenger traffic ceased on 19 September 1936; and goods and mineral traffic ceased on 2 June 1937. In August 1937 the dismantling of the railway began; and in February 1944 the Welsh Highland (Light Railway) Company was wound up.[14]

Broad-gauge railways reached the Ffestiniog district in October 1867, when the Cambrian Railways opened a new section from Barmouth to Pwllheli via Portmadoc.[15] When exchange sidings at Minffordd were opened in 1872, some slate traffic was diverted from the Festiniog Railway.

The expansion of the slate mines near the village of Ffestiniog was hindered in the second half of the nineteenth century by lack of rail transport. In the 1860s efforts were made to promote a railway from Blaenau Ffestiniog to Ffestiniog, with branch lines leading to the slate mines so that slate could be taken to Portmadoc along the Festiniog Railway.

Samuel Holland and his brother Charles were the leading promoters, and the Festiniog & Blaenau Railway Company was established in 1866. The shareholders included the Holland brothers, J. W. Greaves, other quarry owners, shopkeepers, farmers and innkeepers. The capital was increased from £10,000 to £20,000 in January 1868; and on 30 May 1868 the line was opened. It was 3½ miles long, and its terminal was above Ffestiniog. A branch was constructed to Craig Ddu Quarry, but this was the only quarry it reached. The average gradient was 1 in 66, and most of the track was in cuttings or on embankments. The slate traffic did not develop as had been hoped; but the railway, although it proved a poor investment for the shareholders, was useful for the carriage of passengers and goods.[16]

In 1879 the Festiniog Railway's traffic was further reduced by the extension of the London & North Western Railway to Blaenau Ffestiniog. In 1880 the LNWR joined Llandudno Junction with Duffws, the terminus of the Festiniog Railway; and in 1883 the Bala to Blaenau branch line of the Great Western Railway was opened. The arrival of the Cambrian Railways, the LNWR and the GWR meant that the Ffestiniog area was linked with South Wales, Lancashire and the Midlands. This diverted business from the Festiniog Railway; and it also led to the decline of Portmadoc, and of the small ships which carried the slate coastwise to British ports.

The expansion of motor transport in the 1920s and 1930s resulted in the suspension of winter passenger traffic on the Festiniog Railway; and with the outbreak of World War II, passenger traffic stopped entirely. In 1946 freight trains ceased to run, but the company did not go into liquidation. In 1951 the Festiniog Railway Society was formed, and in 1954 the society acquired more than half of the shares in the company. The line was reopened in stages, most of the work being done by volunteers; and the section between Portmadoc and Minffordd was opened to passenger traffic on 19 May 1956.[17]

In 1870 the Caernarvon to Afonwen section of the LNWR was opened; and this superseded the narrow-gauge railway from Nantlle to Caernarvon. The Corris, Machynlleth & River Dovey Tramroad was authorised on 12 July 1858; it linked slate quarries in the Corris area with the River Dovey at Derwenlas near Machynlleth. It was opened for slate traffic on 30 April 1859, and for passenger traffic between Machynlleth and Corris on 4 July 1883. Passenger traffic was extended to Aberllefenni on 25

August 1887. Passenger services ended on 1 January 1931; and the last train for freight traffic was on 20 August 1948. Since this date the line has been abandoned. The gauge was 2ft 3in; and the section between Machynlleth and Derwenlas was abandoned after the opening of the Cambrian line from Machynlleth to Borth in 1863. The title of the tramroad was changed to the Corris Railway by an Act of 1864 (27 & 28 Vic Cap 225); and this Act also authorised an extension of the Upper Corris branch to Garthgynfawr, which was never made. In 1878 the Corris Railway came under the control of the Imperial Tramways Company Ltd; and then in the same year that company became part of the British Tramways & Carriage Company. The Corris line was prosperous until 1900, and a dividend was paid until 1905. The decline of the slate industry, however, caused a reduction in traffic. The line was taken over by the GWR under an Act of 4 August 1930.

The Corris Railway used road vehicles as feeders to its passenger trains. At first these vehicles were horse wagonettes; but from 1918 they were motorbuses. One service which ran from Corris to Talyllyn Lake and Abergynolwyn played its part in a 'Grand Tour', organised by the Corris Railway, the Cambrian Railways and the Talyllyn Railway. The track of the Corris Railway was taken up in 1949, and the two remaining locomotives were purchased by the Talyllyn Railway Preservation Society in 1951.[18]

The Talyllyn Railway Company was incorporated on 5 July 1865, and was designed from the start for locomotives. It was opened for goods traffic in July 1866, and for passenger traffic in December 1866. The railway was built to link Bryneglwys Slate Quarry with Towyn; and its gauge was 2ft 3in. The engineer was James Swinton Spooner, whose younger brother introduced steam locomotives on the Festiniog Railway. Originally the line had been expected to run from the quarry to Aberdovey; but the authorisation of the Aberystwyth & Welsh Coast Railway meant that the line from Towyn to Aberdovey was given up, and construction of the line was limited to the $6\frac{3}{4}$ miles from Abergwynolwyn to Towyn. There was a mineral extension to the slate quarries at Bryneglwys, which were worked by a firm of Manchester cotton spinners. The gauge was the same as that of the Corris Railway, as the slate quarries at Upper Corris and Bryneglwys were close neighbours. The first steam locomotives were purchased from Fletcher Jennings & Company Ltd of Whitehaven. The first, *Talyllyn*, had an 0–4–0 wheel arrangement, and and second, *Dolgoch*, was an 0–4–0 with a very long wheelbase of 6ft 6in.

The approval of the Board of Trade had to be obtained before passenger services could be started; and an inspection of the railway was therefore carried out on 8 September 1866. The report listed seven bridges over and fifteen under the railway, in addition to a viaduct 51ft high and 38yd long across the Dolgoch ravine. The bridges over the line had 'a span on the square of only 9ft 1½in or 9ft 1in', and as the first passenger coach had a width of 5ft 3½in in outside measurements, this left only 1ft 11in between the outside of the coach and the abutments. On one side of the coaches, therefore, the doors and windows were permanently fastened and barred; and the rails were adjusted so as to give enough space between the other side and the abutments. At the time of this report the only stations on the line were Towyn Pendre and Abergynolwyn; the Wharf Station was used only for transference of slate to the Aberystwyth & Welsh Coast Railway. Later there were six stations: Towyn Wharf, Towyn Pendre, Rhydyronen, Brynglas, Dolgoch and Abergynolwyn.

Bryneglwys Quarry produced 3,816 tons of slate in 1907; but when it was sold in 1911 to Sir Henry Haydn Jones it was in a dangerous condition, many chambers having no adequate roof support. No further capital was invested in the quarry; but in the 1920s the Talyllyn Railway developed as a tourist attraction. After the quarry's closure in 1948 and Haydn Jones' death in 1950, it seemed possible that the railway might cease to operate; but L. T. C. Rolt called a public meeting in Birmingham, and the Talyllyn Railway Preservation Society was formed.[19]

The Glyn Valley Tramway ran from Chirk in Denbighshire to Glyn Ceiriog and Pandy. It was 8¼ miles long, and had a 2ft 4¼in gauge. Its function was to link slate, granite and silica quarries near Glyn Ceiriog and Pandy with the Shropshire Union Canal and the GWR. The Cambrian Slate Company, established in 1857, needed transport for an estimated output of 4,000 tons of slate; and in 1870 they obtained authority to build a line from the Shropshire Union Canal at Chirk Bank to the Cambrian Quarries above Glyn Ceiriog. This horse-worked tramway was opened in April 1873, and passenger services began in April 1874. Steam locomotives were introduced for mineral traffic in July 1888, and for passenger traffic in March 1891. Passenger services ceased in April 1933, and all traffic ceased in July 1935.[20]

The slate quarries around Dinas Mawddwy were served by the Mawddwy Railway. This was of standard gauge, and connected with the Cambrian Railways at Cemmaes. The company was incorporated on 5 July 1865, and the railway was opened on 1 October 1867. The bad condition of the track made it necessary to close the line in April 1908; and it was subsequently taken over by the Cambrian Railways. Having been rebuilt to light railway standards under the supervision of the Cambrian's engineer, the line was reopened on 31 July 1911; but it soon proved to be unprofitable. The Cambrian Railways were taken over by the GWR on 1 January 1923. The Mawddwy line was closed to passengers on 1 January 1931, and to goods on 6 September 1950.[21]

The slate quarries round Hendre Ddu in the Corris district were served by the Hendre Ddu Tramway, which was horse-worked and had a gauge of 1ft 11½in. It connected with the Mawddwy Railway at Aberangell, and it had been closed by 1939.[22] Several slate quarries were served by short mineral lines. The Deeside Slab Works near Corwen were served by a tramway 3 miles long with a gauge of 2ft; the Cwm Ebol Slab Works near Machynlleth by a tramway 1½ miles long with gauges of 1ft 11½in and 3ft on its two inclines; and the Minllyn Quarry near Dinas Mawddwy by a tramway ½ mile long with a gauge of 2ft 2in, which ran to exchange sidings with the Mawddwy Railway.[23]

The development of these railways, which served all the successful slate mines and quarries, led to a decline in the coastal trade in slates. Until about 1880 this coastal trade continued to flourish; but after that date railway competition led to a decline, while wooden schooners were gradually replaced by iron ships. Between 1851 and 1900 the number of ships registered at Caernarvon fell from forty-one to two. Until 1867 the Ffestiniog quarries sent all their slates by sea from Portmadoc; but the situation changed when the area was linked to the main railway lines of northern and central Wales. The percentage of the total slate output sent by sea from Portmadoc fell from 100 in 1867 to under 5 in 1925.

As the foreign markets for slate expanded, it was sent in large vessels from London, Liverpool and Southampton. World War I, however, reduced the prosperity of the shipping industry, and thereafter slates were sent by rail or road to Harwich and then by ferry to the continent.[24]

The Anglesey Shipping Company, which owned a fleet of steamers, was purchased by Lord Penrhyn in 1897; but it continued to be worked as a separate concern. In 1914 seven ships were operating from Port Penrhyn: the *Pandora*, the *Penrhyn*, the *Harrier*, the *Bangor*, the *Linda Blanche*, the *Mary B. Mitchell* and the *Pennant*. The *Mary B. Mitchell*, a three-masted schooner of 380 tons, was requisitioned by the Admiralty in April 1916.[25] During World War I the Admiralty used Port Penrhyn as a base for mine sweepers, and the Penrhyn steamers were kept busy supplying these mine sweepers with coal.[26] In January 1915 the *Linda Blanche*, which was carrying slates, was blown up in the Irish Sea by *U21* ten minutes after the crew had taken to the boats; and in July 1919 the company claimed compensation from the German government.[27] In May 1919 the *Mary B. Mitchell* was sold for £6,500; and in 1921 the company acquired two new vessels, the *Sybil-Mary* and the *Pamela*. The cargo capacity of the company's vessels was as follows:

Pennant	680 tons
Pamela	380 tons
Bangor	330 tons
Penrhyn	330 tons
Sybil-Mary	240 tons

Harrier	180 tons
Pandora	180 tons [28]

Penrhyn Quarry found it unprofitable to use their fleet of steamships; and by 1939 only half the output of slates was being sent by sea. In March 1922 Lord Penrhyn was told by W. Hobson, the quarry manager, that 'Freights to the French Coast from the East Coast and Bristol Channel have fallen from about £5 to 8s 6d, and from the South of England with clay to the Mersey from 50s to 8s 6d'.[29] Increases in seamen's wages also contributed to the decline of sea transport. The *Harrier* and the *Penrhyn* were sold in 1926 and 1937 respectively.[30] World War II virtually ended the slate traffic from Port Penrhyn; and the harbour is now used mainly by small fishing vessels. Penrhyn Quarry does, however, export some slate both from there and from Port Dinorwic.

The tonnage of slate shipped from Port Dinorwic declined steadily from 1897 to 1918:

Year	*Tons*
1897	72,658
1898	74,301
1899	66,934
1900	61,679
1901	69,388
1902	70,530
1903	72,737
1904	69,264
1905	64,208
1906	62,189
1907	59,450
1908	58,263
1909	60,422
1910	56,572
1911	54,219
1912	61,119
1913	55,407

Year	*Tons*
1914	47,145
1915	27,483
1916	15,979
1917	10,899
1918	9,195

The tonnage of slate sent by rail or sea from Caernarvon during these years was approximately half the tonnage shipped from Port Dinorwic. There was a gradual increase, however, in the proportion of slates sent by rail; and in most years after 1900, more slate left Caernarvon by that means than by ship.[31]

Tonnage of Slate Shipped from Caernarvon

	1897	*1898*	*1899*	*1900*
	Tons	*Tons*	*Tons*	*Tons*
Per vessel	26,433	21,498	19,840	14,980
Rail	14,970	17,767	18,982	19,909
Total	41,403	39,265	38,822	34,889
	1901	*1902*	*1903*	*1904*
	Tons	*Tons*	*Tons*	*Tons*
Per vessel	17,921	17,955	16,739	12,963
Rail	21,251	19,969	17,909	17,263
Total	39,172	37,924	34,648	30,226
	1905	*1906*	*1907*	*1908*
	Tons	*Tons*	*Tons*	*Tons*
Per vessel	9,172	10,242	8,353	6,914
Rail	19,731	20,311	16,824	15,167
Total	28,903	30,553	25,177	22,081
	1909	*1910*	*1911*	*1912*
	Tons	*Tons*	*Tons*	*Tons*
Per vessel	14,778	6,875	8,516	8,632
Rail	12,029	13,816	5,220	2,225
Total	26,807	20,691	13,736	10,857

	1913 *Tons*	*1914* *Tons*	*1915* *Tons*	*1916* *Tons*
Per vessel	7,753	6,390	3,322	2,101
Rail	1,789	2,558	3,552	465
Total	9,542	8,948	6,874	2,566

	1917 *Tons*
Per vessel	567
Rail	1,296
Total	1,863 [32]

The fleet of steamers owned by Dinorwic Quarry in 1953 included four vessels: *Juliet Duff*, *Alfred Mason*, *Veronica Tennant* and *Dawlish*. The last three were sold in the 1950s for a sum of £35,250.[33]

9
THE EXPANSION OF HOME AND OVERSEAS MARKETS

During the period of canal building from 1790 to 1830, an expansion of the home market for slate was made possible by reductions in transport costs; and after the Napoleonic Wars the cost of carriage as a proportion of the price of slate was progressively lowered. The cost of sending slate by water from Caernarvon to London, as a percentage of the value of the slate at the port of lading, was 61 per cent in 1790 and only 12 per cent in 1880.[1] This relative decline in transport costs was an important factor influencing the increased demand for slates. Competition between railways and coastal shipping became keen in the 1860s, with the arrival of the Cambrian Railways in North Wales; and this caused a fall in the cost of carriage by water.

The cost of slates rose during the nineteenth century. The cash price for 1,200 best Duchesses at Caernarvon was £7 10s 0d in April 1842.[2] That of Cilgwyn best Duchesses at Caernarvon or Nantlle was £11 in 1887 and £12 17s 6d in 1906.[3] In the case of the coastal markets the decline in transport costs as a proportion of value was due more to the rise in prices than to the fall in freight rates. The cost of shipping slate from North Wales to London in 1880 was only 2s 9d per ton less than in 1790; but the cost of a ton of Countesses at Caernarvon increased by £3 1s 6d during the same period.[4]

The abolition in 1831 of the *ad valorem* tax on slate carried coastwise resulted in a fall in the cost of slating; and slates therefore began to replace tiles in London, Cheshire, Lancashire and the Midlands. Even before the tax was abolished, however, the growing market was producing an increase in the number of specialist slate-brokers. In 1805 Lord Penrhyn made agreements with Samuel Worthington, Michael Humble and Samuel Holland, by which they undertook to dispose of most slates produced at Penrhyn Quarry.[5] The agreement with Samuel Worthington, a Liverpool merchant, continued until 1819.

The lack of respectable slate merchants allowed slates to be hawked from port to port by speculative traders, who were also masters of vessels. Such a system was very unsatisfactory, as is shown by the following representative case.

In March 1816 Peter Roberts, a Liverpool slate-broker, alleged that a man called Ives had asked him to find a purchaser for a cargo of slates. The 'slate trade being at that time depressed', Roberts recommended that the cargo should be sent to Manchester; and this was done. Roberts and Ives met another man in a Manchester public house; and this man agreed to purchase the slates at 47s per ton, if Ives would take as part-payment a debt of £28 7s which was due to him from William Jones of Caernarvon. An agreement was made whereby Ives agreed to sell 38 tons of slate at 47s per ton, and to pay the carriage from Runcorn. Ives was to be paid £20 in cash, £12 2s od in goods, and £28 7s od through transference of the debt; and the purchaser 'accepted for the remainder'.[6] This agreement resulted, not surprisingly, in a legal dispute; but the outcome of the dispute is not recorded.

Often slate merchants proved dishonest. An instance of this was a slate merchant called David Roberts, who was imprisoned at Caernarvon in 1828 as an insolvent debtor. An account of his case was given by Hopkin Perkins of Swansea to a slate merchant of Caernarvon:

> The said David Roberts chartered a Vessell to bring some Slate to this place—and in order to pay the Captain his freight, he clandestinely engaged to purchase a Cargo of Coal—and . . . he promised to pay me before he left the place for the said Cargo. He left this place with considerable property, in his possession . . . he wrote to me from

> Birmingham on his way home that he wo.d remit me for the Coal. The quantity of Coal shipd on Board the *Hope* of Pwllheli for the above was 72 tons at 9s 6d, £34 4s od. I must beg you will act for me to oppose this man—and to have his affairs scrutinised to the utmost. This transaction with me is one of the most daring plunder.[7]

From 1808 until his death in 1817, George Ford acted as the London agent for his brother-in-law William Turner; and his letters show how varied the functions of an agent could be. On 16 June 1808 Ford was endeavouring to send corn to North Wales; but there was 'never a Vessel loading nor can I hear of anything coming to Barmouth or Traeth'. After reporting an order for 12 tons of slabs, he continued: 'The furniture you Ordered shall come by the next Vessel that Loads for your place —and have heard that sister wants some chairs. You get so lusty that you have broken yours pretty well, and they shall come with the other things. I will thank you if it will make no difference to you to put more Countesses and less Ladies as they don't go off quite so well.'

Some of Ford's difficulties were outlined in a letter of 22 June 1808. Freight and timber were high; and vessels were difficult to obtain, as the Government was taking 'all from 80 to 400 Tons at 25s per Ton three months certain'. The war affected the stability of banks and business houses. On 10 July 1810 Ford told his brother-in-law that many 'large Houses in Merkentile Business' were 'stopt'; and he added that 'several other Houses' were 'on the toter'. On 15 January 1811 Ford was in need of slates; and he also asked Turner to buy him a pig 'about the same size as before'. On 19 October 1811 Ford added a PS to his letter: 'I hope you are loading us all you can at Traeth as we are in Want. Your cement and Candles and Soap are in the *Athalia* and Mr. Casson's Candles and Soap are with them.' On 31 March 1814 Ford referred to the 'Bangor Quarry' underselling Turner, and said that the Slate Trade was 'very dull'.[8]

James Wyatt became the agent for Penrhyn Quarry in 1819, when he succeeded his father Benjamin; and he retained this position until the second half of the century. A Penrhyn Quarry letter-book covering the period from 31 January 1848 to 20 December 1858 contains entries which illustrate the marketing arrangements of one of the largest slate concerns. The customers who featured most prominently included William Dawbarn in Liverpool, Richard Cooper in London, George Shelton in Birmingham and Hugh Montgomery in Belfast; and there was a considerable trade with Dublin and Cork. During the years 1848–58 slates were exported to Hamburg, Magdeburg, Trieste, Boston and New Orleans.

Certain principles were laid down by Penrhyn Quarry about the home trade. On 31 January 1848, George Jones Webb of Llanelly Colliery was told that they would be pleased to supply him, but would prefer cash payment at a discount of 2½ per cent. They added: 'We could not take any Coals or other Articles in part payment.' On 4 February 1848 Wyatt declared that payment of the freight was 'quite foreign to our Business'. Four months' credit was allowed; but a 2½ per cent discount was offered for 'payment by bank order on receipt of invoice'.

At the beginning of 1848 freight rates from Bangor were as follows:

London	From 14s od to 15s od per ton, according to the size of the vessel
Southampton	From 13s od to 13s 6d per ton, according to the size of the vessel
Chichester	From 14s 6d to 15s 6d per ton, according to the size of the vessel
Hitchin	From 15s od per ton, according to the size of the vessel
Maldon	From 16s od per ton, according to the size of the vessel
Chester	From 5s 6d per ton, according to the size of the vessel
Runcorn	From 5s od per ton, according to the size of the vessel
Bristol	From 9s 9d per ton, according to the size of the vessel
Liverpool	From 5s 6d to 7s 6d per ton, according to the size of the vessel
Birmingham	From 5s 5d per ton, according to the size of the vessel
Coleraine	From 7s 6d per ton, according to the size of the vessel
Cork	From 10s od to 10s 6d per ton, according to the size of the vessel

Page 189 (*above*) View of a schooner being loaded in 1890 at Greaves's Wharf, Portmadoc; (*below*) view of Greaves's Wharf at Portmadoc, showing the slate trucks

Page 190 (*above*) Port Penrhyn about 1913; (*below*) Caernarvon Slate Quay

Freight rates fluctuated considerably: between February and May 1848 the rate to London fell from 15s 0d to 10s 6d and the rate to Maldon from 16s 0d to 10s 6d. The lower rates, however, were described as 'extraordinary'.

The range of available sizes resulted in local preferences: in the South of England 20in × 10in was the commonest size, while smaller slates were liked in the Midlands. A thicker slate was preferred in Lancashire and in Scotland; and some parts of Ireland used the larger sizes, 24in × 12in and 24in × 14in. During the nineteenth century blue slates were the most popular.

Penrhyn Quarry was always careful to establish the financial reliability of their customers. On 31 January 1848 they told Mr Fawcett Fothergill of Boston, USA, that they gave credit only to 'persons of known means and capital' and that they would prefer to supply him for cash. Crowden & Grange & Company were reassured on 14 February 1848 that the 'ordinary request of being satisfied as to the responsibility of a party' before a credit account was opened was not meant to give offence, and that they had asked no more of the company than they had asked of 'hundreds of others before without any complaint'. One of the many customers about whom inquiries were made was Sir William Jardine: on 31 July 1848 information about him was sought from W. Maxwell of Dumfries 'as a *mere matter* of business (knowing that even Titles are not always exempt from Financial difficulty).' John Murphy of Belfast was told on 21 February 1848 that 'Our Bankers do not like any Bills made payable anywhere but either London, Dublin or Liverpool.'

During the 1840s Penrhyn Quarry did little business with Manchester. In a letter of 1 September 1848 to R. Massey, Wyatt declared it fortunate that they had 'an abundant market elsewhere', since 'we should literally starve if we depended on Manchester.' The quarry dealt only with customers who bought whole cargoes of slate; and such customers existed in many parts of the British Isles. Places in England to which slate was sent included London, Liverpool, Birmingham, Bristol, Chester, Cambridge, Devizes, Bedford, Hitchin, Littlehampton, Chichester, Lewes, Southampton, Alnwick, Portsmouth, Dorchester,

Frodsham, Maldon, Sheffield, Poole, Oxford, Wisbech, St Ives (Hunts), Leighton Buzzard, Salford, Leicester and Bath. Welsh destinations included Bangor, Flint, St Asaph, Amlwch, Pentraeth and Beaumaris; and destinations in Scotland included Hawick, Girvan, Lockerbie, Bo'ness and Dumfries. Irish orders were sent to Londonderry, Belfast, Sligo, Dublin and Letterkenny; but the last of these ports was not 'well-liked' by Bangor shipmasters.

As a rule Penrhyn Quarry did not employ agents to sell their slates; but they had an agent in Liverpool. On 11 December 1851 Wyatt informed Isaac Taylor that their yard in Liverpool was opposite the Princess Dock Basin; slates for the Liverpool area could be supplied from that yard through W. Dawbarn and Company.

Arrangements with William Dawbarn in Liverpool had been reviewed in April 1850; and Wyatt wrote to Dawbarn, informing him that the results for Colonel Pennant were not satisfactory. Over the nine years of Colonel Pennant's management of Penrhyn Quarry, average exports through the agency had been £4,142 a year and average commission £207 a year. The average home sales were about £3,584 a year, with a profit to Colonel Pennant of about £81 a year; and this left 'a balance in the Agent's favor of £126 a year besides their half profits of £81 and commission in the Home Sales of about £180 a year'. A new arrangement was put forward, whereby Dawbarn remained agent for all slates except writing slates and hones. All profit on the home sales was to go to him; and the commission on export sales was to be at 4 per cent. The trade in writing slates and hones was to be conducted by Dawbarn independently and not as agent.[9]

The 'Hungry Forties' affected the slate industry seriously. In February 1842, for example, a letter from a quarryman appeared in the *Caernarvon and Denbigh Herald*. It referred to the general depression in 'trade, manufactures and shipping throughout the Kingdom . . . Building is nearly put a stop to, or greatly checked. This causes there to be a very little demand for slates; and we poor quarrymen shall soon feel still more the sad effects. We shall be thrown out of employment, or our wages much reduced.'

At the same date 'one of the people' complained about the '14 or 15 loads of oatmeal' sent from Anglesey 'to feed Mr. Assheton Smith's hounds.' The correspondent thought it would have been 'much more meritorious and charitable if this article had been given to the distressed and half-starving poor in the neighbourhood.'[10]

The railway boom of the late 1840s, however, stimulated new building, and thus increased the demand for slate. On 12 July 1848 Wyatt wrote to Rupert Brindley in London regretting the depression in trade. He contended, however, that this depression would be short-lived: 'as soon as people feel confident they will set to work building again.' Ten days later he told George Shelton in Birmingham that the quarry's slates were in good demand, and that they had no intention 'at present' of making a reduction.

Complaints about underselling were voiced in 1848 by Assheton Smith's customers in London. On 13 June Wyatt wrote to Richard Cooper, saying that these customers found Penrhyn slates being retailed for less than they themselves paid on the wholesale market. He told Cooper that as there was an understanding between Colonel Pennant and Mr Smith that there should be no underselling of each other, he had investigated the matter. He had discovered Cooper's prices for Countesses 'quoted in your *own handwriting* at £6 5s od per thousand per cash'. This was about 7s per thousand below the cost price in London, which was made up as follows:

Countesses at Bangor	£5	2s	6d
Freight at 16s per ton	£1	2s	od
Stowage and straw		1s	4d
Insurance		1s	6d
Unloading		1s	od
Lighting		1s	6d
Landing		1s	od
Wharf rent and loading costs		1s	6d
Total	£6	12s	4d

On 8 April 1858, in a letter to N. Matthews of Tremadoc, Wyatt made further inquiries about underselling. He had been informed that 'Port Madoc best Slates' had been 'offered at 10 per cent under our prices in the English and Bristol Channels.'

On 18 April 1851 Wyatt made an arrangement with William Davey of Bath, whereby Davey was allowed a commission of 5 per cent on all best quality roofing slates from the size of 20in × 10in upwards. The brick duty had been abolished in 1850 and trade was now improving steadily. Davey was told that 'our sales since January have considerably increased particularly in London or the East coast and English Channel.' Wyatt continued: 'If other parties think to gain upon us by competition and underselling or other means, they will be disappointed in the end for if we are burthened with stock of any kind the only plan is to lower away in prices, wages etc. I am however inclined to think myself that we have seen the worst. I do not think *any* of our neighbours can increase; it will be well if they can maintain their supply. The rest must depend on the demand.' Wyatt referred sceptically to the new mode of transport: 'You are very much afraid of the Railways. I do not like it myself, but there is no help but to go on with the stream or be stranded, but I think their powers are limited as regards slate.'

It was generally believed that the Hamburg fire of May 1842, which destroyed public buildings and over 2,000 houses, gave the Welsh slate industry its first opportunity to export slates to Germany. In March 1848 the freight rate from Port Penrhyn to Hamburg was from 16s to 18s per ton; and terms of payment were 'a satisfactory guarantee from a House in England'. In 1848 slates were also sent to customers in Magdeburg; but exports were affected by the political events of that year. Writing in March to Josti & Company of Magdeburg, Wyatt said he was 'sorry to hear of the alarming state of things on the Continent'. On 8 April 1848 Wyatt thanked this company for settling their account. His letter continued: 'These are indeed awful times, and it is impossible to see what will be the end of it, but I hope there is enough of good sense in Germany and elsewhere, if there is not in France, amongst the body of the people to preserve peace and order rather than involve everything in anarchy and ruin.' On 7 April Wyatt told Captain Matthews of Tremadoc that he wished 'we could punish this rascally Frenchman for disturbing the peace of Europe.' On 28 June Wyatt told W. H. A. Voigh of Hamburg that the situation in Europe was 'still very bad'; but he hoped that 'a favourable change' would allow the company 'to renew . . . Shipments to the Continent.' On 28 July Wyatt wrote to Henry Harris & Son in Poole: 'It is impossible in these times to say what may happen. We may have a revolution or a civil war before anyone would suppose it, seeing how quickly they do these things on the Continent.' In fact, however, the United Kingdom experienced neither a revolution nor a civil war in 1848.

The American trade of Penrhyn Quarry was mainly concentrated on Boston and New Orleans. The freight rates to Boston in March 1848 were '6 to 7 dollars per ton'. The slates most popular in the American market were 'best blue ladies', which were 16in × 10in and 16in × 8in. In February 1848 a customer was warned that 'it might be two months before a vessel could be loaded'; and it was explained that the quarry was 'shipping largely to America'. As events on the continent complicated trade relations, so the American market became more important. On 3 March 1852, however, a customer in Boston was told that his cargo of 16in × 10in slates could not be sent for 122 days. Delays were also caused by the limitations of sailing ships. On 5 August 1848 Wyatt wrote to a customer in Cavan: 'I am very sorry you are kept so long without your slates, but you must see that we cannot possibly help it nor can any human being control the elements.'[11]

Until the 1840s the trade in Ffestiniog slates was mainly coastal; but there were also markets in Ireland, including Dublin, Drogheda, Belfast, Londonderry, Dundalk, Waterford and Cork. The ports in Great Britain which traded most extensively with Portmadoc were London, Hull, Liverpool, Chester, Stranraer, Ayr, Bristol, Gloucester, Runcorn, Gosport, Leith, Glasgow, Aberdeen, Sunderland, South Shields, Hartlepool, Swansea, Cardiff, Newport and Southampton.[12] In the 1840s Portmadoc began to export slates to Hamburg and the Baltic. The quantity exported annually was about 10,000 tons, or one-fifth of the total shipments. Most of this trade passed through Hamburg; from there slates were forwarded by river, canal and railway to many places in Germany and Austria. Thousands of tons were exported annually to the Baltic ports, including Copenhagen, Malmö, Stettin, Danzig and Königsberg. Cargoes were also sent to Rotterdam, Dunkirk and Calais.

The expansion of the Australian cities provided Britain with another market for slates; and large quantities were exported from Portmadoc to Melbourne. Australian houses had generally been roofed with timber; but slates were found more satisfactory, and in some cases they were also used on the walls. From 1852 to 1854 large quantities of slates were sent from Portmadoc to Liverpool and then transhipped into Australian liners. By 1856, however, the Australian markets were glutted; and there was a temporary decline in exports to them. Slates were also shipped via Liverpool to America, including California; to the East and West Indies; and to the Cape of Good Hope. The tonnage of slates shipped from Portmadoc rose from 11,396 in 1825 to 46,802 in 1855.[13]

The Great Exhibition of 1851 ushered in a period of prosperity for Great Britain. New capital was invested in industry and agriculture; public transport was expanded; and industrial production suffered no serious disruption from strikes or lock-outs. The growth of joint-stock banking provided a sounder foundation for Britain's credit system; and the discount rate of the Bank of England, which acted as lender of last resort, controlled the rates for short-term loans both in Britain and elsewhere.[14]

The Exhibition stimulated the demand for Ffestiniog slates from foreign countries. Many firms won prizes, and John W. Greaves was awarded the prize medal. In 1854 Samuel Holland sent specimens of his slates to the Paris Exhibition, and was given honourable mention there.[15] Hopes were then expressed that the heavy duty imposed on slates imported into France from Britain would be repealed; and in 1860 the Cobden Treaty with France was concluded. This treaty gave a stimulus to the export trade, as

France had hitherto been a closed market. A deputation of North Wales quarry owners, including J. W. Greaves and S. Holland, asked the President of the Board of Trade to include slates in the commercial agreement. The duty had previously been from 4s 9d to 30s in the pound on slates; but after the treaty, the duty was from 8d to 10d per ton. No great expansion of trade took place, however, as the mines and quarries of Angers and the Ardennes could supply the French market more cheaply. The treaty was followed in the 1860s by similar agreements with Norway, Sweden and Denmark; and America and the Netherlands also reduced their duties. These changes caused a remarkable expansion in the export of slates to Europe, especially as the development of the continental railway system gave the Welsh slate producers access to inland markets. The value of Welsh slates exported to western Europe increased from £55,000 in 1856 to £155,000 in 1880.[16]

Germany was the chief importer of Ffestiniog slates; and until the end of the nineteenth century there was no serious competition from other slate-producers, or from substitutes. The Oakeley Company's Cash Book for 1868–78 shows that their principal continental markets were Frankfurt, Hamburg, Cologne, Flensburg, and Copenhagen.[17] At this period Oakeley exported no slates to America. The American trade had been affected by the Civil War of 1861–4; and despite a brief post-war revival it never regained its former importance.

Until the arrival of the Cambrian Railways, all the slates of the Ffestiniog district were shipped from Portmadoc. Portmadoc slates were popular abroad because of their lightness of weight and uniformity of colour; but on the home market they were less highly valued. On 26 August 1884 the secretary of the Llanberis Slate Company wrote to J. P. Clarke in Cheapside: 'Portmadoc slates are altogether a different class of Slates to ours and very inferior as a rule.'[18] The Ffestiniog district was thus more dependent upon the foreign market than were other areas of slate production; and its trade was badly hit by the Crimean War, the American Civil War, and the Franco-German War of 1870–1. The value of slates exported fell from £192,576 in 1869 to

£160,853 in 1870; but by 1873 it had risen, as a consequence of rebuilding, to £271,986.[19]

Between 1876 and 1880 the value of slates exported fell by more than one-third. This was the result of a fall in commodity prices, which was experienced throughout the Western world and led to a stagnation of trade known as the 'Great Depression'. That phrase, however, is misleading. The three recessions of 1879, 1886 and 1894 were separated by the two booms of 1883 and 1890.[20] The peak year for the export of slates was in fact 1889, when the figures reached 79,900 tons and £278,840. The general recovery of trade after 1894 led to a brief revival in the slate industry but this was succeeded by a steady decline, and by 1914 the quantity of slate exported had fallen to 24,214 tons.[21]

10
THE GROWTH OF UNIONISM

The quarrymen of the Ffestiniog district were said in February 1873 to have made themselves 'less notorious than their brethren the colliers in South Wales. They never in this district go on strike.' With minor exceptions later, this remained true; and industrial relations never became as bitter as in Penrhyn, Dinorwic, and some of the smaller Caernarvonshire quarries. One reason for this greater contentment was a generally higher level of wages. In 1873 the Ffestiniog quarrymen were receiving from £1 10s to £2 weekly; and the labourers who removed the waste rock were paid on average £1 5s per week.

The slate-workers of this area were regarded as a 'very orderly set of men, not given to heavy drinking and debauchery', 'Almost without exception', they were Liberal in politics and Nonconformist in religion. The quarry owners did not interfere to any great extent in the men's lives; and they included the Whig Lord Palmerston and the Liberal Samuel Holland. The quarrymen were a scattered group: most of them did not reside in Blaenau Ffestiniog but lodged there during the week, returning home at weekends. About 120 lived in Trawsfynydd; and the arrival of standard-gauge railways in the 1870s enabled these workers to travel daily. In 1873 there were 'upwards of twenty' chapels in the Ffestiniog district; and there were only sixteen public houses. Tanygrisiau had no public houses; and Samuel Holland, the chief landowner, was said to be against them.[1]

In Penrhyn Quarry and in Dinorwic Quarry a paternalistic control of the workers was in force; and this feudal relationship was broken only by the impact of World War I. These two slate quarries, the largest slate quarries in the world, were regarded as wonders not to be missed by travellers. In 1852 Richard Edwards informed tourists visiting the Penrhyn Slate Quarry that he was appointed guide from the Waterloo Inn, Bethesda, to the Devil's Kitchen and Carnedd Llewelyn.[2] The influence of the Pennants and the Assheton Smiths on Caernarvonshire was very extensive during the first half of the nineteenth century; and between 1832 and 1868, except for one parliament, the county seat was held by one or other of them. As landowners they were able to keep their tenants subservient by threats of eviction and by the non-renewal of leases; and as industrialists they controlled by the end of the century a total work-force of over 5,000. The following verses from the Llandegai School Log Book for January 1868 give some indication of the prevailing atmosphere:

Then welcome merry Christmas
And the blooming Christmas tree,
Bless Lord and Lady Penrhyn,
And may they happy be
And bless the kind young ladies
Whom little children love,
And may they live for ever,
In the blest world above.[3]

The Pennant family regarded their tenants and work-people as objects of charity; and the family enjoyed giving these dependents small treats from time to time. In December 1846, for example, 'Lady Louisa Pennant distributed in person, clothing and shoes to the boys and girls (shoes to the former; shawls, cloaks and shoes to the latter) attending Llandegai schools, upwards of two hundred in number.'[4] Another popular treat was an organised trip; and a novel rail-outing was arranged on 29 June 1850. 'About 800 of quarrymen from the works of Colonel Pennant, together with wives and sweethearts, visited Chester on Friday by means of the Chester & Holyhead Railway.'[5]

The Pennants and the Assheton Smiths, like many other quarry owners, were divided from their workers not only by class but also by language and religion: most of the quarrymen spoke Welsh, and the majority were Nonconformists. Before 1865, however, these differences did not result in serious conflict, even in Penrhyn and Dinorwic Quarries; and it would be wrong to assume that they made such conflict inevitable. There were in any case many factors which reduced the likelihood of industrial disputes: the geographical barriers separating the various industrial communities; the quarrymen's local rather than national patriotism; the concentration of interest on the chapel rather than on politics; the smallholdings which gave an illusion of independence; and the fact that many employers combined genuine benevolence with hostility towards any combination of labour.

Even before 1865, however, there were serious grievances. Reference has already been made to the 1825 dispute at Penrhyn over inequalities in wages, which ended with a return to work and the punishment of the men's leaders. In the early 1840s the slate trade was depressed, and 300 workers were dismissed from Penrhyn Quarry. The average wage at Penrhyn in 1845 was 15s per week for most of the workers;[6] and in 1846 many quarrymen emigrated to the USA, since they were unable to find employment at home. A strike was attempted in Penrhyn Quarry in 1846; but it proved a complete disaster. After a message had been passed round asking the men to strike, one of the overseers ordered them back to work; and only result was that the carrier of the message was punished by being demoted to work as a rubbisher. The trouble had arisen from the same causes as in 1825: disparity of earnings, favouritism and low rates of pay.[7] Wages varied within a single quarry, because of differences not in skill but in the quality of rock allotted to various groups. There were also variations in day rates.

In April 1845 men working by the day at Cwmorthin Quarry in Ffestiniog were earning the following amounts:

	Day's rate	£	s	d
John Evans	24 at 2s	2	8	0
Richard Jones	23 at 1s 6d	1	14	6
Cad. Thomas	24 at 1s	1	4	0
William Thomas	12 at 2s	1	4	0
Ellis Williams	11 at 2s	1	2	0

Under this piece-rate system it was almost impossible for the quarrymen to know what the average wage was; and without such a point of reference it was difficult to establish that a grievance existed. The system known as poundage was established in all quarries by the 1830s: this meant that an advance was paid on every pound's worth of slates produced, the slates being valued at the standard list price for making. At Cwmorthin Quarry in 1844–5 the list prices for the best quality slates were 30s per 1,000 for Princesses 24in × 14in, 27s 6d per thousand for Duchesses 24in × 12in, 20s per thousand for Countesses 20in × 10in, and 11s per thousand for Ladies 16in × 10in. Poundage in one case was 20s; and a total wage for making 3,780 slates was £2 18s 6d.

Managers were often aware of the worker's financial difficulties, whereas quarry owners were conscious only of the need for economies. Henry McKellar, the London director of Cwmorthin, writing to the manager, Allen Searell, on 25 August 1843, impressed on him the 'necessity of getting the work at the quarry done on the most reasonable terms—labour, I understand from a neighbour of yours, should be done just now *very moderate*—but that we do not perceive in your account or cost sheet. Pray see carefully to this.' Searell replied on 28 August that the men had earned less for their work in June and July than was earned at other quarries. He gave an example of the days worked in two bargains:

> Thomas Kemp and Company, 210 days. The total amount of their bargain is £17 1s 6d, less powder, ironsmith etc, £5 11s 6d; real earnings, £11 10s 0d, which will not give

1s $1\frac{3}{4}$d a day, and Thomas Kemp has to pay some of his men that are working with him by the day 2s 6d per day.

Hugh Morgan and Company, 155 days. The total amount of their bargain is £9 0s 3d, less iron, tools etc, 5s 6d, real earnings, £8 14s 9d, which will at best be little over 1s $1\frac{1}{2}$d per day. Ever since I have had anything to do with the works I never knew men work so hard as they have and earn so little money. There is not a good Quarry Man to be got in the neighbourhood for less than 2s 6d per day and not a first rate one for that and when they take a bargain they expect to get more but these have worked six weeks for almost nothing after paying their men and cost.[8]

Between 1846 and 1865 there was discontent with conditions in many of the larger quarries. The bargain system had become a disruptive force. A group or partnership of workers was allotted a section of the quarry; and in the larger quarries the group sold the slates to middlemen or contractors, to whom parts of the quarry were sub-let. At the end of a month a wage settlement was made, and the money was shared out among the partnership. The contractors were often blamed for carrying out the policy of the owners, who were usually isolated from their workers; and a system of bribery and corruption prevailed, because the level of wages was so low that men resorted to dishonest practices in order to make a bare living.

In the 1860s prosperity had come to the slate industry, and wages had risen. The table of annual wages for a quarryman at Dinorwic from 1851 to 1860 shows that there had been a steady increase:

	£	s	d
1851	35	3	4
1852	40	19	10
1853	40	16	10
1854	54	1	5
1855	54	18	11
1856	54	16	4

	£	s	d
1857	55	2	4
1858	55	12	6
1859	53	15	0
1860	55	5	2 [9]

After the repeal of the Combination Acts in 1825, the movement towards the formation of Trade Unions began gradually to gather strength. The Amalgamated Society of Engineers was established in 1851; and it served as a model for new unions combining all sections of one industry. In North Wales the 1850s were free of strikes; but in England the disputes of that decade included an engineers' strike and lock-out in 1852, a strike of spinners in Preston and other Lancashire towns in 1853, and a strike of coal-miners in Yorkshire in 1858. In 1859–60 the building operatives of London went on strike to obtain a nine-hour day. This was followed by a lock-out; and the Master Builders' Central Association attempted to force the operatives to sign what was known as the 'Odious Document'. This was a written guarantee that the operative would not belong to any society which interfered in any way with arrangements made between the employer and his workmen.[10] A similar policy was later to be adopted by the owners of Dinorwic and Penrhyn Quarries. The quarry communities were isolated; but the men could read about English labour struggles in such newspapers as the *Herald Gymraeg*.

At Penrhyn Quarry in the spring of 1865 a committee was formed by Robert Parry, Robert Thomas, Richard Evans and Richard Pritchard. These men sent a letter to Colonel Pennant asking him to remedy their wages complaint; and on 31 July, when no reply had been received, the men came out on strike. Robert Parry and his colleagues asked William John Parry (1842–1927) if he would go with them to meet Colonel Edward Pennant. As 'Quarryman's Champion', W. J. Parry was to exert a considerable influence on the growth of unionism in the slate quarries, and on the progress of events from this date until 1903. Thereafter his importance declined; and by the time of his death his popularity had been reduced by his own intolerance. He was the son of a quarryman in Penrhyn Quarry; but because of his father's ownership of two houses, he was able in 1854 to attend the Llanrwst Grammar School, where the instruction was in English. In 1857 he was employed by a Bangor solicitor named H. Lloyd Jones. By the time of the 1865 strike he had become an Associate of the Society of Auditors of England.[11]

After meetings with Colonel Pennant, at which W. J. Parry acted as interpreter, concessions were granted; and the men resumed work after a strike lasting fourteen days. The success of this strike action encouraged the men to form a union, and they asked W. J. Parry to draw up a plan.[12] The scheme was made public in September, and 1,800 workers immediately joined. Quarry owners everywhere were alarmed; and on 21 November 1865 William Hawkins, one of the London directors of the Welsh Slate Company, wrote a letter on the subject to Samuel Holland. Having heard that the 'protective union by the Quarrymen at Bangor' was 'intended to include all Wales', Hawkins wondered what steps should be taken. 'Will you therefore kindly let us know what course you think it will be best for us to adopt? Although a Trade Union such as we have in London would be an enormous Evil in our quiet district, still it does not appear to me clear that a Union of any kind would be certain to reach our Neighbourhood.' In his reply of 22 November, Samuel Holland said: 'I cannot learn that anything has yet been done in this Neighbourhood by the Quarrymen towards adopting the suggestions thrown out by the Bethesda Men—and I think until we have correct information that a move is taking place among our Men—we had better not let them suppose that we fear it. When we learn that they are really contemplating the formation of a Union here, then we must decide what steps to take to check it.'[13] These letters show very clearly that Holland objected to unionism even when it was qualified by the adjective 'protective'.

On 2 December 1865 Colonel Pennant issued a notice which strongly condemned the attempt 'by certain self-sufficient individuals, calling themselves a committee, to form a species of Trade Union among the Quarrymen.' Colonel Pennant advised his employees not to listen to this committee, 'the effect of whose proceedings . . . would be to estrange him from his quarrymen, towards whom he has always hitherto endeavoured to act as a friend, and to establish an ill feeling between himself and his workmen.' Colonel Pennant called on the older quarrymen to tell their younger colleagues of the improvements made in the previous twenty-five years 'as regards cottages, gardens, clubs, schools, hospitals'. He cautioned the men against having anything to do with 'agitators' or with a trade union, and continued: 'On the very first rumour of such a state of feeling, he will immediately close the quarry, and only re-open it and his cottages to those men who declare themselves averse to any such scheme as a Trade Union.'

On 22 December 1865, 1,229 of the quarrymen replied to these threats by asserting that the committee was made up not of 'agitators' but of delegates appointed by a majority of the quarrymen. The men continued, however: 'We also beg to state that we have entirely renounced the idea of a Trade Union, trusting that Colonel Pennant will act according to his principles that there will be no more "severity and revenge".'[14]

In the Parliamentary Election of November 1868 Colonel Pennant's son George Douglas Pennant lost his seat for the county of Caernarvon to the Liberal candidate, T. Love D. Jones-Parry. The results were as follows:

T. L. D. Jones-Parry	1,963
George Sholto Douglas Pennant	1,815

The numbers polled in Bethesda and Bangor were:

	Parry	*Pennant*
Bethesda	47	312
Bangor	111	258

Jones-Parry had addressed himself particularly to the quarrymen; and in his election address of 28 October 1868 he had urged them to fight against 'bondage' and 'tyranny'. Pennant, on the

Page 207 (*above*) Penrhyn Castle about 1850; (*below*) Lord Penrhyn with the Penrhyn Quarry Committee in 1874

Page 208 (*above*) Dinorwic Quarry, showing a slate block being unloaded by crane from a lorry; (*below*) Penrhyn Quarry, showing quarrymen using pneumatic drills

other hand, had expressed the fear that a part of his constituency was being 'dazzled by the specious promises and words of persons, who, for their own purposes, agitate the public mind.'[15]

At the time of the 1865 strike, W. J. Parry had been associated with a company which supplied Penrhyn Quarry with gunpowder; but on 15 July 1868 he told F. C. Wickson that he was unlikely to receive any other further order, as the quarry manager John Francis was personally hostile to him. Parry attributed this loss of business not to his action during the strike but to a letter he had written for one of the quarrymen in 1867. This letter had contained a complaint about John Francis; and on seeing Parry's handwriting Francis had taken offence.

The strike of 1865 resulted in the quarrymen at Penrhyn Quarry having their monthly wages raised by about 30s; but corruption gradually returned to the quarry, particularly after George Pennant's electoral defeat. On 20 March 1869 W. J. Parry took up the case of four quarrymen who were being victimised for supporting the Liberal candidate: two of them had been dismissed and two had been demoted to day work. Parry expressed the view that Edward Pennant, who became Lord Penrhyn in 1866, must be unaware of the situation; but he received no answer to his letter.[16]

By the beginning of the 1870s labour relations were deteriorating in most of the quarries, with the exception of those centred on Blaenau Ffestiniog. Wages in Penrhyn Quarry were very low, while profits were consistently high. Wages were 24s a week on average, while in other quarries they averaged about 30s. In 1870, because the industry was depressed, eighty workmen were dismissed from Penrhyn Quarry. Most of the men chosen had been involved in the 1865 strike, in the setting up of a trade union, or in the 1868 election at which Pennant lost his seat. On 13 December 1873 the *Caernarvon and Denbigh Herald* referred to 'trouble brewing in the slate districts of North Wales'; and it went on to show that the workmen's position in the Caernarvonshire quarries was 'simply wretched'. The newspaper gave a typical four-week budget for a quarryman's household consisting of two adults and five children:

	£	s	d
Rent at 2s 6d per week		10	0
Bread	2	0	0
Coals		12	0
Meat		8	0
Potatoes, sack		7	0
Clothing		12	0
Butter 3lb at 20d per week	1	0	0
Milk		2	0
Sugar 3lb at 4d per week		4	0
Tea 1½lb		4	6
Candles 1½lb at 4d per week		2	0
	6	1	6

This was a minimum and did not include crockery or furniture; and the writer challenged quarry owners to deny that many workmen in the quarries were paid £3 and less per month. Slate prices had risen over the previous decade, the writer argued, but wages had not.[17] An answer to these claims was published the following week. It stated that the first writer was guilty of exaggeration; that many quarrymen had small farms or kept lodgers; and that not one quarry in a hundred resembled those of 'the late Lord Palmerston and the present Lord Penrhyn'.[18]

There was great discontent at Dinorwic Quarry in 1874. The *Caernarvon and Denbigh Herald* attributed this unrest partly to the owner, George William Duff Assheton Smith, who had succeeded to the estate in 1869.[19] The former owner had been Thomas Assheton Smith, the great fox hunter, and it was maintained that the 'old gentleman thought more of his "white jackets" (quarrymen) than even of his greyhounds, not withstanding his eccentric enthusiasm.' The men, it was said, admired the character of their new master; but they disliked his way of organising the quarry.[20]

In January 1874 a movement was begun to establish a trade union at Dinorwic Quarry. Some of the quarrymen involved with the 1865 committee had suffered for their actions; and so it was agreed that officers should be chosen from men working outside the quarries. W. J. Parry was chosen as secretary; and Morgan Richards, a Bangor tradesman who later wrote a book on slate quarrying, was chosen as president. Hugh Pugh, a Caernarvon bank manager, was the treasurer.[21]

The organisation was at first called the Society for the Defence of Slate Quarrymen; but it was later renamed the North Wales Quarrymen's Union. On 17 July 1874 Lord Penrhyn complained to his solicitor about the 'mischief' which was being caused by this body. Referring to the county of Caernarvonshire, he declared that such events made him 'despair of any permanent good being effected in it'; and he concluded, 'I have done my best for thirty-four years and I now give up for ever.'[22] In fact, however, he retained control of Penrhyn Quarry until 1884; and when he died in 1885 his place was taken by his son, George Douglas Pennant.

On hearing of the setting up of a trade union, the employers held a meeting at the Royal Hotel, Caernarvon. All the major Caernarvonshire undertakings were represented: Penrhyn, Dinorwic, Glynrhonwy, Penybryn, Cilgwyn, Dorothea, Cambrian, Coedmadog, Talysarn, Snowdon Slate Quarries, Moeltryfan, Vronheulog, Fron, Braich and Cloddfa'r Lon. The representatives issued a circular declaring 'that in the opinion of this meeting every quarry proprietor in North Wales should refuse to employ any man who is ascertained to be a member of the union, and that every proprietor or his representative should give notice of such determination to his men at the earliest opportunity.'[23] War had been declared; but Lord Penrhyn disapproved of the role played by his manager at this meeting and instructed him to write to the newspapers to say that he had not been consulted.[24]

The first incident occurred at Glynrhonwy Quarry, which lay on the opposite side of Llyn Padarn from Dinorwic. The managing director, Captain Wallace Cragg, was informed that 110 of his 128 quarrymen had joined the union; and on 21 May 1874 he told them that they must withdraw from the union or leave his employment. When they refused to withdraw, he closed the quarry;

but the men had no difficulty in finding work elsewhere, since the demand for slate was at that date high. The lock-out ended three weeks later; and the men were taken back as unionists.

The action of Captain Cragg was followed by Assheton Smith's notice to his quarrymen that on 18 June 1874 Dinorwic Quarry would be closed unless the union was disbanded. Many hundreds of his workmen had joined the union, and a deputation of the men met Assheton Smith. He was ready to listen to their grievances: the unsettled conditions at the works since the death of the chief agent Henry Turner; the appointment of Colonel Wyatt as chief manager; and the appointment of an English auditor who by his new rules about time keeping had caused 'the greatest irritation to the men'. He refused, however, to tolerate a union;[25] and this refusal was the main reason for the lock-out which followed.

All but eleven of the Dinorwic Quarrymen gave their allegiance to the union; and so for 'the first time within the memory of men now living', the 'great Dinorwic Quarries' came to a standstill. The prosperity of the slate industry at this period removed some of the drama from the event. An account of the lock-out reported the immediate departure of the Anglesey labourers from the quarry: 'The moment they heard the threat of a lock-out, these Anglesey labourers went to their different barracks, took up even their beds . . . The young quarrymen seemed very unconcerned . . . scores of them left the neighbourhood for Nantlle, Ffestiniog, Corris, and other places, where they have already obtained employment. The labourers are going to the hay-harvest, and many of them are determined to seek work upon the railways and not return to the quarry.' The same account reported a rumour that the proprietors of the many smaller quarries hoped to profit from the stoppage. Such men, it was alleged, were urging the 'young squire to crush the union by locking out his men'; but they themselves were careful 'not even to protest against their workmen joining that society'.[26]

After a stoppage of five weeks, the men at Dinorwic Quarry returned to work. Both sides had made concessions. The management agreed to accept the union; but the union's rules were modified by the inclusion of the two new clauses. The first of these stated, 'That this Union has no intention to interfere with the managers, agents or working of the quarry, or any of the rights of the proprietors'; and the second stated, 'That no workman who will not join the Union is to be molested on that account.'[27]

The quarrymen of Penrhyn Quarry had made a collection of £206 in support of the men locked-out from Dinorwic; and on 14 July Lord Penrhyn reacted to this news with the following notice:

> Being informed that a large body of the workmen in the Penrhyn Quarries had given support to an Union formed at Llanberis for the purpose of dictating to the owners and managers how their quarries should be worked, I hereby give notice that I shall resist any such interference with the rights of proprietors, and shall, if such support be continued, immediately close the quarry.[28]

This notice was regarded as a challenge, and the Penrhyn workmen hastened to join the union. Penrhyn Lodge was established on 20 July with almost 2,200 members; and on 31 July 1874 the men came out on strike.[29] They had at least one serious grievance, in that their wages were lower than those paid in other quarries. On 9 September 1874 W. J. Parry appealed for funds to support the strike, saying:

> Over 2,200 Quarrymen are out on strike and above 300 old men have been 'turned out' by Lord Penrhyn, for whom some provision must be made to keep them from the Union House. Some of the old men have been working at Lord Penrhyn's Quarries 40, 50 and 60 years, and some of them even more than this, and simply because the bulk of the Quarrymen ventured to ask for an increase of Wages, these poor innocent old men were cruelly punished by being thrown upon the parish for any provision Lord Penrhyn was ready to make for them. This certainly is not charity.

> The men advanced three grounds for asking for an increase, viz. The state of the slate market, the advance in the price of the necessaries of life, and the wages allowed in other Quarries. Against these three, Lord Penrhyn advanced three for not granting the demands, viz., that granting the demands, the Quarry would not pay what he considered to be proper profit for him, that the men forgot that he was very charitable to them and he could not continue his charities and proper wages and that granting the demands of the men would bring ruin upon themselves and families. No comment is required on these.[30]

The agent of the Penrhyn estate who negotiated with the men's committee was Captain Pennant Iremonger, who in 1872 had changed his name to Pennant Lloyd; and it was to him that the committee wrote on 22 August: 'We are perfectly willing that his lordship should keep his charities to himself—if those in any way interfere with him in his giving us proper wages. We ask no charity, but proper wages for a proper day's work. If our wages are reduced because of these schools, hospitals, and cottages, we do not see them in the light of charities at all, but part of our just wages. We have also to remark that in the way of low rent, cottages, and land, we believe that it will be found that the Llanberis men in this respect are better off than us, and certainly in the way of better wages they are greatly in advance of us.'

During the strike many letters were exchanged among the men, Pennant Lloyd and Lord Penrhyn; but the men were in the position, thanks to the prosperity of the times, of being able to dictate terms to their employer. The final outcome of the negotiations was what was known, to the later embarrassment of the estate agent, as the Pennant Lloyd agreement. This document was considered comparable to the Bill of Rights; but it proved to be dependent on the goodwill of the employer, and on prosperity in the slate industry.

The terms of the agreement are summarised as follows by W. J. Parry:

1 The following prices for working slates:

Blocks		4s	a ton
Writing Slate Blocks		3s	,, ,,
Best Ton Slates		7s	,, ,,
Seconds		5s 6d	,, ,,
Queens		10s	,, ,,
Fourth Tons		5s	,, ,,
Moss Slates		5s 3d	,, ,,
Best 24 × 14	£1	2s 6d	per thousand
,, 24 × 12	£1	8s 4d	,, ,,
,, 22 × 12	£1	5s 10d	,, ,,
,, 20 × 10	£1	0s 0d	,, ,,
,, 18 × 10		18s 4d	,, ,,
,, 16 × 10		15s 0d	,, ,,
,, 16 × 8		11s 0d	,, ,,
,, 14 × 12		11s 0d	,, ,,
,, 14 × 8		8s 0d	,, ,,
,, 13 × 10		9s 2d	,, ,,
,, 13 × 7		5s 10d	,, ,,
,, 12 × 6		5s 6d	,, ,,
Seconds 24 × 12	£1	5s 0d	,, ,,
,, 20 × 10		18s 4d	,, ,,
,, 16 × 10		12s 11d	,, ,,
,, 21 × 12	£1	0s 10d	,, ,,

2 The fixed maximum scale of wages of quarrymen was abolished.

3 The wages of masons were fixed at from 3s 6d to 4s a day, according to merit; and up to 4s 6d, if those wages were granted at Dinorwic Quarries.

4 That platelayers' wages be from 15s to 18s per week, according to merit; and up to 21s, if those wages were granted at Dinorwic quarries.

5 That the wages be paid every four weeks.

6 That a supreme manager and umpire be appointed with powers to decide all disputes.

7 That a Committee be appointed to manage the Penrhyn Quarries Sick Benefit Club, to consist of Lord Penrhyn as

president; the supreme manager as vice-president; a treasurer, a secretary, and one workman from each district in the quarry.

8 That the power to turn men out of the works be in the hands of the supreme manager.

9 That the power to take men into the works be in the hands of the chief working manager, under the direction of the supreme manager.

10 That the right to stop men until their case is inquired into by the supreme manager be in the hands of the chief working manager.

11 That the workmen who are working on rocks at 10s in the £ are to be taken into bargains as circumstances permit; and that the places that are being worked by them now, that can be so let, be let as regular bargains.

12 That the same terms be allowed rubble men as are allowed at the Dinorwic Quarries.

13 That one month be allowed all workmen to return to their places in the quarry.

14 When necessary that partners be allowed to name their partners to be placed before the chief working manager, and if refused by him that the name be placed before the supreme manager, whose decision will be final.

15 That if a quarryman by extra work makes 35s a week, that the agent is not to reduce his price if the rock is of the same quality next month.

16 That the bad rockman be similarly placed at 24s 6d a week.

17 That all complaints about letting be first referred to a committee appointed by the workmen; and if considered by them to be a proper case that it be placed before the supreme manager for his decision.

18 That in the event of a quarryman failing to earn 27s 6d a week for two consecutive months, he is entitled to lay his case before the Committee, and through them before the supreme manager for his decision.[31]

On September 1874, after these concessions had been granted by Lord Penrhyn, the men went back to work. The newspapers announced that the strike was 'happily at an end' after 'six anxious weeks'. The men were praised for showing no disrespect for Lord Penrhyn; and it was said that 'not a finger was laid on the men who continued to work during the strike.'[32] These rejoicings, however, proved premature. The men went on strike again almost immediately, declaring that the managers had totally disregarded the contract and conditions were worse than before. By this time Pennant Lloyd, thinking the matter settled, had left for the continent.

Complaints were therefore laid before Arthur Wyatt, the shipping agent at Port Penrhyn; and the men asked for two arbitrators to decide about the charges made against the three managers, J. Francis, R. Morris, and O. P. Jones. The two arbitrators gave their decisions on 4 November 1874; and in all but one of the eighteen cases these decisions were in favour of the men. On 5 November the men responded by approving the following motion:

> That this meeting, after hearing Mr. Wyatt's report and decisions read, resolve to return to our work under the arrangement come to with Mr. Pennant Lloyd and to take the lettings from others than the present managers.

The men returned to work on 9 November, and the three unpopular managers resigned. J. J. Evans, the manager of Dorothea Quarry, was appointed chief working manager at Penrhyn Quarry.[33] The Penrhyn strike of 1874 thus ended in a victory for unionism. This outcome reflected the improved status of trade unions, which had come about largely as a result of their increased membership. The changes in the law made in 1871–5 gave trade unions freedom to pursue trade disputes without the application of the doctrine of criminal conspiracy; and this allowed picketing to take place. Employer and employed were put on an equal footing in the courts by the law relating to the finishing of contracts; and trade union funds were given the legal protection already enjoyed by the funds of friendly societies.[34]

At Penrhyn Quarry the years 1874–85 were harmonious ones; and during that period Lord Penrhyn's relations with his workmen, and with W. J. Parry, were friendly and happy. By 1878 the effects of a reduction in the demand for slate were being felt. W. J. Parry, however, attributed the disruption of harmony not to economic factors, but to the 'spirit of persecution' introduced early in 1885, when the quarry was taken over by George Pennant.[35]

11
SOCIAL CONDITIONS

The wages system in the slate industry of North Wales was based on piece-work, and rates rose fairly steadily in the second half of the nineteenth century. In 1843 'good quarrymen' were being paid about 2s 6d a day; but by 1865 the rate had risen to an average of 3s 6d a day.[1] At the Oakeley Quarry in 1888 the average pay for a 'slatesman' was 5s a day; miners received 4s 6d a day and 'clearers' 3s 8d.[2] As, after 1850, prices were rising, the increase in wages in real terms was not very great; and in quarries such as Penrhyn the men were often worse off, as their pay failed to keep pace with rising prices.

A comparison of wages at Dorothea and Cilgwyn Quarries on 30 July 1897 shows very slight variations in the rate paid:

	Dorothea	*Cilgwyn*
Quarrymen working by the day	4s 3d to 4s 6d	4s 3d to 4s 6d
Labourers " " " "	3s 8d	3s 6d to 3s 9d
Labourers constantly employed	4s 0d, 3s 10d 3s 9d, 3s 8d	3s 6d to 3s 9d
Plate Layers	3s 10d	4s 0d, 3s 10d 3s 8d, 3s 6d
Draggers	3s 10d	4s 0d, 3s 10d 3s 8d, 3s 6d
Engine Drivers	4s 3d	4s 6d, 4s 3d 4s 0d
Joiners	4s 10d, 4s 5d 4s 3d	4s 3d, 4s 6d 3s 9d
Masons	5s 0d	4s 9d, 4s 3d
Fitters	4s 6d	4s 6d
Weighing Machine Men	3s 2d	3s 3d, 2s 9d
Foreman at Shed	4s 3d	
Pumpers	4s 7d	
Joiners and Pumpers		5s 0d

On 10 August 1899 the manager of the Dorothea Slate Quarry wrote to the manager of Cilgwyn, explaining that they had made a 'general reduction in the prices of Bargain men and labourers working on contracts but nothing in any of the Daymen's wages'. He said that the reduction in the prices of slates was 'telling heavily upon us here—and the reduction we made was fully justified in consequence'. It was not, he added, 'compensating the Company for the reduction in the Slates'.[3]

Wages at Penrhyn Quarry were given much publicity towards the end of the nineteenth century, when labour relations became more embittered. In order to correct the 'many misrepresentations in the Press as to Lord Penrhyn paying his men starvation wages', it was reported in 1885 that the net average daily wage of the quarrymen was 5s 11d and that this rate was higher than in most other quarries.[4]

In 1900 monthly wages at Dorothea Quarry were as follows:

	Number of Men	*Wages* £	s	d
Quarrymen	246	5	8	6
Rubbishers	83	2	12	3
Labourers	79	4	10	5
Horses	10	7	6	8
Daymen	59	4	11	8
Contractors	14	5	10	4
Slate Loaders	6	6	10	3
Slate Inspectors	8	6	18	2
Total	495			

The depression in the slate trade in the early twentieth century resulted in unemployment and the reduction of wages. In 1910 the monthly wages were as shown on page 221.

Besides taking away the waste rock, the rubbisher carried material from the working-places of the rockmen or quarrymen to the place where the slates were made.[6] Another term in common use was 'rubbler' or 'rybelwr'; and every quarryman had to pass through this first stage of apprenticeship. When a boy had his name entered on the quarry register, he decided on a certain gallery in which to work; this was usually one where a member of his family was already working. He was given a small block of slate; and he was shown how to trim it, and how to split and dress the blocks. If a boy was unlucky he might spend the rest of his life begging for blocks; but the majority became first journeymen and then bargain-men or members of a crew.[7]

	Number of Men	*Wages*		
		£	s	d
Quarrymen	126	4	15	5
Rubbishers	57	2	12	3
Labourers	41	4	1	6
Horses	6	6	16	4
Daymen	43	4	10	5
Contractors	3	3	5	1
Slate Loaders	6	3	4	8
Slate Inspectors	7	6	5	11 [5]

The rockmen and quarrymen were the men who entered the bargain system. By various modes of drilling and blasting, the rockmen brought down large blocks of slate and then split them into smaller blocks. These blocks were sent to the quarrymen who worked in the slate mills and split the blocks into slates.

A description of pay-day in Dinorwic Quarry, printed by the *Slate Trade Gazette* in 1914, showed that a regular ritual was observed. A bugler announced that the money was ready; and at 10 am it was brought by special train from the nearest bank. After its arrival, office staff had to count out the money into small, round, tin boxes, which were stacked on the floor. At the signal from the steam hooter all the men surrounded the office; and after a bell was rung, the tin boxes were thrust into the the quarrymen's hands, whereupon little groups could be seen intently counting their money.[8]

Efforts to obtain a minimum wage were made in 1911 by the North Wales Quarrymen's Union. They drew up a charter, which regulated hours of work and holidays and fixed a minimum daily wage of 4s to quarrymen, rockmen and other skilled workmen, such as blacksmiths and masons. Miners were to receive 3s 9d; rock labourers 3s 6d; and labourers 3s 4d. The minimum wage was to be paid weekly, not monthly; and no deductions of any kind were to be made.[9] The employers for the most part rejected the idea of a minimum wage; and the Minimum Wage Bill failed to get a second reading in 1914.[10] In 1918, however, a minimum wage was conceded at the daily rate of 4s 6d for quarrymen and 4s for labourers. In October 1918 a sliding scale was drawn up to vary with the selling-price of slates. The minimum day rate for quarrymen was 5s 9d; for rock labourers 5s 6d; for day men 6s 6d; and for labourers 5s 9d.[11]

In 1894 it stated that it was difficult to find any working man who had more leisure time than the quarryman of Caernarvonshire. The hours in Penrhyn Quarry were given as from 6.30 am till 5.30 pm, and on Saturdays from 6.30 am till 11.30 am, but in winter the hours were shorter. The dinner hour from 11.30 to 12.30 was usually spent reading books or newspapers in the cabins, or in preparing for the Eisteddfod.[12] Hours varied slightly from quarry to quarry; and there were sometimes differences between official and actual hours. On 4 May 1895 the manager of Oakeley wrote to the manager of Dorothea Quarry: 'We allow half an hour all the year round for dinner, but in as much as we know that this cannot be done we say nothing on seeing the men leaving their respective working places before 12 provided the procedure is not abused. We all leave off at 2pm the first Monday of the month.'[13] In 1919 the working hours were reduced from 52½ to 47½ hours a week during eight months of the year. The new working hours for the rest of the year were 44 hours for two months and 41½ hours for the other two months.[14]

Emigration by quarrymen to the USA had begun in the eighteenth century; and during the 1840s the numbers increased, because of unemployment and poor conditions in the quarries of Caernarvonshire and Merionethshire. Between 1795 and 1860 a total of 1,146 persons left the Ffestiniog district for the USA.[15] Emigrants went mainly to New York State; but other places of settlement were Pennsylvania, Wisconsin, and Ohio. R. B. Pritchard left Caernarvon in the early 1870s. On 11 September 1871 he wrote to his brother, 'This is the Country for money to a working man'; but he considered the laws 'disgusting', since there were murders committed every day in New York and 'not one out of every twenty punished'. Pritchard successfully organised his own slate company, the Eagle Red Slate Company, Middle Granville, New York State; and by the 1890s it was competing against the Welsh slate producers.[16]

Information about emigration to South America was given by Walter Holland to Samuel Holland on 28 April 1875; and a depressing picture of the primitive conditions in Argentina was drawn. The scheme to acquire emigrants cheaply was described as follows:

> The Buenos Aires Government has recently imposed an iniquitous tax on all regular steamers bound to their Country, by obliging them to carry free on each voyage six steerage passengers. The object is to encourage emigration at the expense of the steamers. The consequence is that we (among others) are now compelled to carry and feed all kinds of people that offer themselves to the agents of the Buenos Aires Government in this country. These people consist for the most part of Welsh men and Women desirous to get out to the Welsh Colony at Chaput in the Argentine Territory. The other day we had to carry out a large family of Pews (or Pws) from Ffestiniog. We have now to take out another family of Williams and Jones from Tan y Grisiau, no doubt tenants of yours and among these are four quarrymen. One of the Welshmen told me that there were 500 of them to go out. If we have to carry them all it will involve a loss to us of over £5,000. You will also lose the use of many of the men at the quarries. Can nothing be done to stop these infatuated people from going out to such a miserable place? It occurs to me to put some statement of the lawless condition of the country to which they consign themselves in the Ffestiniog Papers . . . but if this were traced to us or even to you . . . it

> would bring down 'hot lead' upon our heads . . . If you hear of people going out it ought to be sufficient to point out to them that everyone is leaving the country that can find the means of doing so and that the labour market out there is quite overstocked.[17]

In 1879 W. J. Parry visited the USA on behalf of the North Wales Quarrymen's Union to discover the conditions and prospects in the slate quarries of Pennsylvania, New York and Vermont; and although he reported unfavourably, the union assisted quarrymen who wished to emigrate to the USA and to the colonies.[18]

The depression in the slate industry resulted in 'hundreds' emigrating in 1906 from Blaenau Ffestiniog to the USA, where they hoped to find employment in the American slate quarries. In addition, about 200 quarrymen from the Ffestiniog district went to South Wales; and others had gone to Australia and Patagonia.[19] In September 1908 'painful scenes' were enacted in the Ffestiniog district 'owing to the departure of batches of slate quarrymen to the coalfields in South Africa'. Work was so scarce that hundreds had left the area.[20]

During World War I a number of families left the quarrying districts for the South Wales and Lancashire coalfields; and during World War II, when many slate quarries and mines were closed down, many men went with their families to work in Wrexham, Birmingham and Port Ellesmere.[21]

In the early years of the slate industry the rate of immigration was greater than that of emigration. At the end of the eighteenth century, a large number of workmen came from Anglesey; and others came from Ruthin, Rhos (near Wrexham), Llanrwst and Llandudno. The quarries of Penrhyn and Dinorwic were popular with the immigrants, not because of higher wages but because of such attractions as pensions, houses, and benefit clubs. The 1841 census figures for the main quarrying parishes show that the Bethesda district (ie the parishes of Llanllechid and Llandegai) had the highest number of people born outside the area. The Llanberis district (ie the parishes of Llanberis,

Llanddeiniolen and Llanrug) came next; and the Nantlle area (ie the parishes of Llanllyfni, Llanwnda and Llandwrog) came last. The figures were as follows:

	Number of persons	*Those born outside the area*
Bethesda area	7,967	1,030
Llanberis area	6,986	685
Nantlle area	6,281	423 [22]

Immigration was considerable in the parish of Ffestiniog. From 1851 to 1861 the increase in population was 1,093. This represented a percentage increase of 31·59, of which 14·31 was natural increase and 17·28 immigration.[23]

Between the years 1801 and 1881 there was a remarkable growth of population in the areas of slate production. Population increased by 350 per cent in the Bethesda district, 600 per cent in the Llanberis district, 300 per cent in the Nantlle district, and 1,400 per cent in the parish of Ffestiniog. Between 1861 and 1881 the number of quarrymen in Caernarvonshire increased from 6,464 to 8,408; and in the same period the number of quarrymen in Merioneth increased from 1,215 to 4,265.

Year	*Bethesda area*	*Llanberis area*	*Nantlle area*
1801	2,602	1,819	2,873
1821	4,205	2,987	4,072
1841	7,967	6,986	6,291
1861	10,627	9,240	6,847
1881	11,878	12,994	11,841

Year	*Ffestiniog area*	*Percentage increase*
1801	732	
1821	1,168	56
1841	3,138	94
1861	4,553	29
1881	11,274	53 [24]

Thereafter the population in these areas began to decline:

Year	*Bethesda area*	*Llanberis area*	*Nantlle area*
1901	9,222	12,069	12,115
1921	7,635	9,709	9,693
1951	7,398	9,385	9,316

Year	*Ffestiniog area*	*Percentage reduction*
1901	11,435	6
1921	8,138	21
1951	6,920	6 [25]

By 1901 the parish of Bethesda had come into existence.

Bethesda dated from 1820, and was named after an Independent Chapel there. In addition to this chapel, it contained an inn and a group of quarrymen's cottages; and other chapels were soon built. A quarry village often developed around a chapel; and such villages often took the biblical names of these institutions. Bethel, Ebenezer, Carmel, Saron and Nebo were industrial communities of a new kind.[26]

Most of Bethesda was built on the Cefnfaes estate, where building leases (usually for thirty years) were granted in the 1830s. Most of these leases were granted to quarrymen; and the estate increased in value from £80 a year around 1830 to £1,700 a year in 1888.[27] In 1854 an Act (17 and 18 Vic 1854 Cap CXI) was passed relating to Bethesda and its neighbourhood. There were clauses dealing with paving, cleansing, lighting and sewers; and the owners of courts and passages were required to flag and drain them. The villages of Gerlan and Llanllechid, like Bethesda, were built on land outside the boundaries of the Penrhyn estate. The Pennants were the chief landlords of the Bethesda area, however; and Llandegai was a model village created on their land at the end of the eighteenth century. The quarry villages of Mynydd Llandegai and Tregarth grew up during the 1840s and 1850s. At the time of the Royal Commission on Land in Wales and Monmouthshire in 1894, Lord Penrhyn owned about 72,000 acres; and the Penrhyn estate included areas in Lleyn, in

Eifionydd and in the parishes of Llandegai, Llanrhychwyn, Llanllechid, Llanddeiniolen, Bangor, Aber, Llanfairisgaer, Llanbeblig, Llandwrog, Caerhun, Gyffyn, Llanbedr, Yspytty, Penmachno and Llanwrst.[28]

The Pennants were commonly regarded as good landlords; and in January 1870 a letter in the *Caernarvon and Denbigh Herald* described Lord Penrhyn's part in the development of Gerlan, a 'pretty new village about half a mile distant from Bethesda'. The village had been designed in 1864–5 in accordance with the 'stringent' rules of the Bethesda Board of Health; and this, it was claimed, had made it 'one of the finest villages in Wales'. Lord Penrhyn had paid £300 for the drainage tanks; and although the village was not on his estate, he had paid another debt of £550. Lord Penrhyn, the writer concluded, was 'one of the most liberal of earthly lords that we know'.[29] The Act of 1854 had created the Bethesda Improvement Commissioners, and Lord Penrhyn was the first chairman. He had contributed £1,000 for the water supply, lighting and drainage schemes. Before these improvements were introduced, outbreaks of typhoid and typhus fever had been common.

Population figures from 1801 to 1881 for the parishes of Llandegai, Bangor and Llanllechid showed the following increases:

Year	*Llandegai*	*Llanllechid*	*Bangor*
1801	1,280	1,322	1,770
1811	1,770	1,470	2,383
1821	2,341	1,964	3,579
1831	2,600	3,075	4,751
1841	3,010	4,957	7,232
1851	3,398	5,948	9,564
1861	3,381	7,346	10,662
1871	3,393	7,739	10,825
1881	3,587	8,291	11,370

Thereafter the population began to decline in the parishes of Llandegai and Llanllechid; but there was no comparable decline in Bangor, which was not so dependent on the slate industry for employment.[30]

In the Llanberis district the Assheton Smiths were the most important landowners. Like the Pennants they gave out plots of about eight to fifteen acres at nominal rents to some of their 'most deserving' quarrymen, on the understanding that the tenants would build cottages for themselves. By 1859 about 2,000 acres had been brought under cultivation by these tenants. The acreage of the Vaynol estate in 1893 was 35,936 acres; and the income from it was £20,394 per annum. In 1893 the agent for the Vaynol estate, Captain Niel Patrick Stewart, explained the rule for the quarry cottages:

> These cottages and quarry holdings have been made especially for the quarrymen working at Dinorwic, and we like to see them kept by the quarrymen, so that when the husband dies we like to see the widow take in a son or nephew, or if that cannot be done, a lodger, in order to make each cottage the home of a quarryman.[31]

The Hon William E. Sackville West, the agent of the Penrhyn estate, stated in 1893 that there was little demand for cottages for agricultural purposes but a 'very great demand for cottages for the quarry'. He said that about 1,600 of the 2,500 quarry employees lived on the Penrhyn estate; and he reported that this estate contained about 873 cottages, exclusive of those in Bangor. He went on to say that with hardly an exception, these had gardens of from one-tenth to a quarter of an acre each. These were chiefly occupied by quarrymen; and there were 130 garden allotments, which were rented at 5s to 10s a year.

Sackville West gave a brief outline of the history of the early cottages and smallholdings. The allotments were of three and four acres each, where one or two cows could be kept. These were of benefit if the quarryman was unemployed; and 'lately' Sackville West had offered the men new allotments with cottages, for about £8 a year. The early cottages had been built with an outlay of about £25, the landlord having provided the slates; and the original leases were for thirty years. When the leases expired, the rents were increased from 10s a year to £3 or £3 10s a year; but such increases were needed to pay for extensive repairs, which cost about £2,000.[32] In the 1890s work in the slate industry was more attractive than agricultural work. In his evidence to the Royal Commission Colonel Charles A. Wynne Finch said that several labourers on his Denbighshire estate had refused to work in the neighbourhood. They preferred to go 6 miles to the Penmacho quarries or 16 miles to those near Ffestiniog. They sometimes returned for the weekend, leaving their families there for the rest of the week. The attraction of higher rates of wages in the slate industry made the labourers reluctant to work on the farm, so they left their families behind in 'comfortable' farm cottages.[33] The wages of an agricultural labourer on the Penrhyn estate were 16s to 18s a week; and this was below the average rates in the slate quarries. Farmers did not migrate so readily, and the agent of the Vaynol estate had heard of no migration of farmers from his district. He had heard of 'a few quarrymen going to America, but many of them return'.[34]

The quarrymen of Dinorwic conducted a correspondence with Captain Stewart in 1876 over the question of the leases of the quarry tenements built by the men themselves on the mountain wastes of Llanddeiniolen. In particular the discussion concerned the question of leases, which were no longer being granted, and the question of compensation for improvements, which the tenants claimed had been allowed them in the days of Thomas Assheton Smith. On 19 May 1876, the men wrote that they were anxious not to 'overstep their proper place as Tenants nor to manifest any rebellious or obnoxious temper or spirit'. They accepted the loss of their early privileges; but they expressed the hope that leases would be granted to them in view of the fact that they had converted a 'wild Mountain into a fine Estate'.[35]

On 16 December 1876 Captain Stewart replied to the men in hostile terms:

> You refer in glowing terms to the administration of this estate in the days of the late Mr. Assheton Smith, and the

> sad contrast present times present—You seem to forget however that real changes have occurred since those palmy days and that the memorialists themselves with their unions and their strikes, have been mainly instrumental in bringing about these unhappy changes . . . Should the memorialists however be pleased to give their servitude *now* on the same terms as then, I am sure Mr. Assheton Smith would be only too delighted to revive and continue the same priviledges and terms of tenure as were granted by his late Uncle.[36]

In 1893 Captain Stewart said there was no demand for leases, but that if there were they would be readily granted. Sackville West in the same year made a similar statement about leases on the Penrhyn estate.[37]

Llanberis, Port Dinorwic and other villages in Llanddeiniolen developed slowly in the 1840s. The ground on which the village of Llanberis was built belonged to the Trustees of the Ruthin Charities; and when first let the land was rocky, inaccessible and of little agricultural value. The original leases were granted in the 1840s to four different persons for sixty years at a 'reasonable' ground rent to the charity. These lessees sub-let the land to quarrymen, who blasted away rocks in order to build houses, borrowing the capital chiefly from building societies such as the Snowdon Permanent Building Society. In many cases the quarrymen paid excessive rents; and only short leases were available. In 1910 some of the leases were due to expire in three or four years' time; and the owners had been served with notices to put their property in good order before the expiration. The owners were prepared to do this, provided they were granted an extension of the lease at an equitable rent. The matter was taken up by the chairman of Llanberis Parish Council in March 1910 with the Hon D. Lloyd George.[38]

The question of the leases involved Edward Foulkes, one of the Dinorwic Quarry officials, and William Jones, MP for the Arfon Division from 1895 till his death in 1915. William Jones received details of housing conditions in Llanberis from Edward Foulkes in 1898. Foulkes described Llanberis as 'an industrial village' whose inhabitants consisted almost entirely of quarrymen. It was not a market-town, and most of the trade was done in Caernarvon. Workmen's houses were much in demand, and many of them were inhabited by two families. The scarcity of houses in the immediate neighbourhood of the quarries forced hundreds of workmen to live at a distance from their work; and they travelled on the Dinorwic Quarry private railway.[39] In 1906 another letter to William Jones explained how '20 or 30 years ago' Llanberis had 'promised to become an important resort for visitors'; but the village had not been able to compete with the seaside towns made accessible by the railways.[40]

The population growth of Llanberis and Llanddeiniolen from 1801 to 1881, due almost entirely to the Llanberis quarries, was as follows:

Year	*Llanberis*	*Llanddeiniolen*
1801	464	1,039
1811	438	1,253
1821	472	1,727
1831	725	2,610
1841	1,024	4,202
1851	1,111	4,894
1861	1,364	5,747
1871	2,507	6,574
1881	3,033	6,886 [41]

In Nantlle the land was owned by many private proprietors who were not concerned with the slate industry. Between 1860 and 1890 the villages of Talysarn, Penygroes and Llanllyfni grew rapidly. All these villages were in the parish of Llanllyfni, which contained the Dorothea Quarry. The population figures for this parish were as follows:

Year	*Llanllyfni*
1801	872
1811	1,128
1821	1,182

Year	*Llanllyfni*
1831	1,571
1841	2,017
1851	2,010
1861	2,362
1871	4,013
1881	5,520 [42]

Of the quarrymen in Caernarvonshire in 1882, 35 per cent were employed in the Llanberis district, 32 per cent in the Bethesda district, 28 per cent in the Nantlle district, 3 per cent in the Conway district, and 1 per cent in each of the Gwyrfai and Pennant valleys. The Caernarvonshire slate industry was thus concentrated on the three chief districts, all within close reach of each other. The improvements in transport made it more difficult for districts far from the coast to compete; and when they were linked by railway it was too late, as the end of the prosperity had by then begun.[43]

In Merionethshire slate production was concentrated in the Ffestiniog district. The majority of the workmen in the slate industry lived in Blaenau Ffestiniog; but before the railway system was established, many lived further away in Trawsfynydd, Garndolbenmaen and Beddgelert. Samuel Holland built many cottages on his estate in the Tanygrisiau district; and later the Oakeley Quarry Company owned a great many houses in the same area. Those living outside the Ffestiniog district were boarded in barracks from Monday to Saturday. Barracks were a feature of the slate industry, and they existed at most of the slate quarries and mines. At Dinorwic barrack accommodation was given free to about 420 quarrymen; and at Rhiwbach Quarry in 1903 the company provided small barracks of two to three bedrooms at a monthly rent of 2s 4d to 5s 8d.[44] In 1895 the doctors' report on the Merionethshire slate mines was unanimous in condemning the barracks then in existence; the faults mentioned included overcrowding, dirt, and the absence of proper sanitary arrangements. The bylaws of the Oakeley Quarry barracks emphasised the need for cleanliness. The rooms were to be 'washed when and as often as the Manager may order'; and every occupier had to keep his bed 'perfectly clean and make the same before leaving for his work, and under no circumstances be allowed to use or keep dirty clothing in his room.'[45]

In January 1874 Samuel Holland wrote to Charles Spooner asking him if he would approve of daily workmen's trains on the Festiniog Railway. Holland had received a memorial from about 120 quarrymen who resided in the neighbourhood of Llanfrothen, Maentwrog and Portmadoc, asking for a daily train to take them to and from their work.[46] Such a train was provided; and it was used by many workmen who lived west of Blaenau Ffestiniog. The development of broad-gauge railways meant that those living to the north of Blaenau Ffestiniog could also travel daily, and no longer had to stay in barracks during the week. In 1893 1,100 workers at the Oakeley mines lived in Blaenau Ffestiniog; and 450 men travelled daily by train.[47]

The growth of the parishes of Ffestiniog and Ynys Cynhaiarn (containing Portmadoc) can be seen from the following figures for the years 1801 to 1881:

Year	*Ffestiniog*	*Ynys Cynhaiarn*
1801	732	525
1811	961	889
1821	1,168	885
1831	1,648	1,075
1841	3,138	1,888
1851	3,460	2,347
1861	4,553	3,138
1871	8,055	4,367
1881	11,274	5,506 [48]

Self-help was practised among the quarrymen; but in times of unemployment, they were forced to ask for poor relief. To prevent this, many leases of quarries (eg Cilgwyn) contained clauses stipulating that a certain number of quarrymen (thirty at Cilgwyn in 1816) should be employed for at least nine months of the year.[49]

Most of the profitable quarries paid large sums of money for poor relief, which in North Wales took the form of outdoor relief, and continued to do so even after the passing of the Poor Law Amendment Act of 1834. During strikes and lock-outs men either lived on credit or left the district in search of work; and in November 1874, after the strike at Penrhyn Quarry, it was reported in the press that only fifteen quarrymen had applied for parochial relief.[50] Another kind of self-help was afforded by local building societies; these became numerous in the 1820s, the quarrymen usually paying a monthly subscription. The Ffestiniog Building Society, established in 1836, was of great benefit to quarrymen who were building their own homes. Sums of almost £2,000 were paid into this society during the first three years of its existence.[51] Another important society was the Merionethshire Building Society, established in 1877, with its registered office at Blaenau Ffestiniog.[52] Many of the houses at Blaenau Ffestiniog, however, were poorly built, being constructed on peat where the surrounding land was waterlogged. Others were built on the edges of the rocks.[53] In 1865 there were as many as twenty-five building societies in Bethesda; and over £18,000 was subscribed to them in a year.[54] The peak of these building societies was reached between 1861 and 1881, when the industry was at its most prosperous. The societies played an important part in the lives of the quarrymen. In some cases their interest rates were too high; but they at least enabled the men to acquire dwelling-houses, and thus gain the right to vote.

The health of the quarrymen improved, as better drainage and purer water supplies in the quarry villages and towns lessened the frequency of outbreaks of typhoid and typhus fevers. The introduction of a pure water supply in the Ffestiniog district reduced the numbers of deaths from these fevers from 12·69 per annum in the years 1865–74 to 1·3 per annum in 1880–90.[55] Overcrowding and poverty made the quarrymen susceptible to other diseases, particularly pneumonia and tuberculosis. As machinery was introduced in the mine and quarries, the danger of 'dust disease' or silicosis was increased; and this often resulted in tuberculosis. It was said in 1902 that many of the quarrymen liked the old methods of sawing, splitting and dressing the slate.

The small huts, built by the men themselves, had often been cold and damp; but the men had been happier and healthier without the noise and increased dust produced by the machinery.[56] The introduction of mechanical drills increased the amount of dust at the rock-face; and the operation of sawing the slabs into blocks caused a considerable increase in the amount of dust in the sheds. Some quarry owners resisted proposals that the alleged high mortality rate from tuberculosis of the respiratory system should be investigated.

A link with the inhalation of slate dust was suspected in the late nineteenth century; but on 12 November 1893 Dr M. Mills Roberts, Surgeon to the Dinorwic Hospital, reported to W. W. Vivian, manager of Dinorwic Quarries, that his visit to Caernarvon to investigate the matter was 'not of much service'; respiratory disease 'among women being 449; among males, not quarrymen at 395; against 506 quarrymen'. Roberts concluded that the matter was 'best left alone'. On 27 June 1918 Thomas Lloyd Williams, wrote to W. Hobson, the manager of Penrhyn Quarry, enclosing a cutting from the *Liverpool Courier*. Williams observed that Ellis Davies, MP for Caernarvon, was 'trying to make out' that tuberculosis amongst quarrymen was caused by slate dust; and he commented as follows:

> I think some years ago this was threshed out and disproved. I further understand from returns that the percentage of Tubercular ailment is greater amongst the Agricultural Labourers of Lleyn than amongst the Quarrymen, and that it is quite as high if not higher throughout the population of North Wales than it is amongst the quarrymen. I fail to see in the face of this why Slate Dust should be responsible for producing phthisis amongst slate quarry workers . . . I trust something can be done to stem this movement, as it would be a very serious matter for the Slate Quarries' Proprietors, if this prevalent disease is appended to the Workmen's Compensation, which the Quarries are not responsible for. It is a matter for Medical Experts to prove whether slate dust and especially the small quantity

the men inhale in their employment, is responsible for producing this disease.[57]

In 1927 Dr T. W. Wade investigated the alleged high mortality rate from tuberculosis of the respiratory system among slate workers in the Gwyrfai Rural District of Caernarvonshire. After analysing the mortality figures, he stated that there was little doubt that workers in slate sheds were subjected to an injurious dust. Further reports on the subject were produced in 1930 and 1933. A tuberculosis physician of the Welsh National Memorial Association published in 1939 a report on eighty-five workers in the Ffestiniog slate mines, whose doctors had referred them to him as having symptoms associated with pulmonary tuberculosis. The results of X-ray examinations showed silicosis in most of the cases. Most of the men were rockmen and miners, but there were some who worked in the sheds. In 1939, as a consequence of this inquiry, the slate mines were included under the silicosis scheme of the Workers' Compensation Acts.

In 1944 an investigation was carried out into the incidence of silicosis among men working in open quarries. X-ray examinations of volunteers in the first quarry visited produced much evidence of pneumoconiosis and silicosis in the sheds; but rockmen in the open quarries were almost completely free from these diseases. The investigators therefore made further examinations of men working inside in other quarries; and more cases were discovered. As a result shed-workers in open quarries were included under the silicosis scheme of the Workmen's Compensation Acts in 1946.[58]

Efforts to minimise the risk of silicosis included water taps fitted over the saws in a large number of quarries; but the resulting wetness of the slabs and floors was unpleasant and unpopular with the men. Respirators were introduced by some quarry owners; but these too were disliked. More success was obtained with extractive fans, which were centrally situated in slate mills; and in 1949 it was proved in Dinorwic and Llechwedd Quarries that these reduced the dust content by 70 to 80 per cent. Dust extractive plants were fitted in the Ffestiniog mines in 1949 and

1950; and these considerably reduced the amount of dust. In Dorothea Quarry water was dribbled on to the sawing machine during rotation, and any residual dust was removed by a system of pipes.[59]

The bad sanitation of the mines in the Ffestiniog district was commented upon in the 1895 report; and it was recommended that closet accommodation should be introduced above and below the ground. The cold damp climate of the Ffestiniog district was blamed by the doctors, in their report of 1895, for the high incidence of rheumatism and diseases of the respiratory organs among the men. Houses were often damp, both from standing on undrained land and from the absence of damp-proof courses.[60] In 1892, however, the manager of Penrhyn Quarry, E. A. Young, claimed that the occupation of a quarryman was as healthy as any outdoor work. Young claimed that the average age at death of Penrhyn quarrymen who died during the decade ending 31 December 1891 had been 'within two and a half months of 59 years'.[61]

The clothing and habits of the quarrymen were often blamed for their poor health. They were accused of drinking too much stewed tea, and of not changing their underclothing often enough.[62] A quarryman at Dinorwic Quarry, giving evidence in 1912, blamed the poor health of the men on the fact that they got wet very often, and were not allowed to go home after getting wet. He said that they were able to get warm in the eating houses, where they provided the coal themselves.[63] Most of the quarries provided eating houses for the men. In 1914 the manager of Dorothea Slate Quarry said that there were 'good shelters' at the bottom of the quarry where the men could take their meats, and that coal was supplied to these places free of charge. There was, in addition, a large dining-room near the saw mills: this was furnished with a stove, tables and benches, where about 200 men could eat. The company employed a man to make a fire, boil water and clean the room; but few of the men used it. The manager claimed that the men preferred taking their meals in their working-places, rather than walk 50yd to the room. In other quarries, however, the 'Caban' where the men took their dinner-break was often important as a meeting-place for debates and competitions.[64]

The health of the workmen was undermined by the frequent occurrence of accidents; and before the Workmen's Compensation Act of 1897 there was little hope of compensation for injuries. In 1842 the Pennants built a small hospital near Penrhyn Quarry for the quarrymen. This example was followed by the Assheton Smiths, who built one overlooking Llyn Padarn; and in 1848 Mrs Oakeley erected a hospital at Llwyn-y-gell for the Ffestiniog quarrymen and their families.

At its foundation, the Oakeley Hospital served other quarries in the area; and the amended rules of January 1866 contained the provision that the hospital was to be open to the men of all quarries in the Ffestiniog district whose proprietors subscribed towards its support. The quarry proprietors subscribed to the hospital in proportion to the number of men and boys they employed respectively. Each workmen was to pay 8d a month, and each boy 4d. A ticket was to be given by the quarry agent to any man who wished to be admitted to the hospital. The medical officer was to be at the hospital every day from 11 am to 1 pm, and was to be available from 6 pm to 8 pm to attend to the outpatients.[65] William Hawkins of the Welsh Slate Company, which subscribed to the hospital, wrote to Samuel Holland in September 1865 about the new rules. Since 'it would be most unwise to offend Mrs. Oakeley, and it would be equally so to attempt to coerce our Men', he proposed that they should deduct 8d from the medical fund 'only from such of the men as might in future use the hospital, and make no deductions from those who might go to other doctors.'[66]

A plea for all the quarries in the Ffestiniog district to use the hospital was made by its surgeon in October 1870. He cited the case of a 'poor man working in Craig Ddu Quarry' who had been carried into the hospital with a shattered hand requiring amputation. Admission had been refused to the man, because the surgeon could obtain no assurance from the Quarry Agent that his company would pay the expenses incurred. The surgeon subsequently attended the man in a 'dark room' about three miles away. The surgeon's report for the year ending October 1870 said that 1,072 casualties had been attended to. During the first three months of the year fever had been prevalent; and several cases of typhoid fever were still under treatment. The draining of the Glanypwll bog had improved conditions; but there was still an 'utter want of main drains in a great many localities in the district'.

In 1870 the Oakeley Hospital fees had been raised to 1s per man per month, with 6d being paid by the boys. Eleven years later the fees were the same, but the hospital was to be used only for accident cases and not for infectious diseases. The fee for any case of labour attended by the hospital's doctor was 10s 6d. The quarries which the hospital served were the following: Llechwedd, Cwmorthin, Diphwys Casson, Rhosydd, and the Oakeley Quarries. In January 1897 a new rule was introduced: 'Hereafter it shall be distinctly understood that this Institution is to be maintained as a Hospital, and not as a Convalescent Home.'[67] By that date the hospital had been taken over completely by the Oakeley Company, while the Llechwedd Quarry had established its own hospital. Other mines had no hospitals, and the workers received hospital treatment only if guarantees were given for payment of their expenses.[68] In 1912 the Royal Commission received evidence about provision for treatment at Dinorwic Quarry. It was said that a doctor treated out-patients free of charge, and that the hospital, like the Penrhyn Quarry Hospital, was also free for the quarrymen.[69] In October 1919, however, it was stated that only accidents were to be attended to at the Dinorwic Quarry Hospital, and that medicine was to be allowed only to in-patients.[70]

Most of the workmen contributed to sick clubs; and the papers of Dr Evan Roberts of Penygroes, quarry doctor in the Nantlle area from 1879 to 1882, show that at least nine quarries made monthly payments of 1s a month to this doctor. The quarries listed were Moeltryfan, Alexandra, Glanrafon, Fron, Dorothea South, Cilgwyn, Cloddfa'r Lon, Coedmadog and Braich. Penyrorsedd Quarry, however, had its own doctor and a first-aid post in the works.[71]

In 1927 the Ffestiniog and District Memorial Hospital was opened; this served the districts of Ffestiniog, Maentwrog, Penrhyndeudraeth, Talsarnau, Harlech, Llanbedr, Trawsfynydd, Dolwyddelan and Penmachno. The establishment of this hospital led to the closure of the slate mine hospitals.[72] In Caernarvonshire the Caernarvonshire and Anglesey Hospital in Bangor replaced the quarry hospitals. These hospitals had improved the conditions of the workmen. The son of the first quarry surgeon at the Penrhyn Quarry Hospital was believed to have been a pioneer in the use of anaesthetics, which he used in 1848 at the quarry hospital for an amputation operation.[73]

Benefit clubs, which gave sickness benefits, were established in most quarries by the 1840s. They were usually modelled on the Penrhyn Slate Quarry Benefit Club, founded in 1787 by Lord Penrhyn and re-established in 1825. In 1825 the subscriptions for this club were 7d per month, and the sickness or accident benefit was 3s 6d a week. The club was organised by the quarry agent and one of the managers. No one over 50 years of age, or in poor health, was allowed to become a member, Subscriptions were deducted from wages at the quarry office, so that membership was in effect compulsory.

By 1874 the Penrhyn Benefit Club had become a subject of controversy. W. J. Parry took up the matter. He brought to light a great many irregularities in the way the club had been functioning, and found a deficiency of over £2,000 in the funds. On 23 January 1875 John Francis was the manager of the club, and he wrote to Parry that coffins were generally paid for by Lord Penrhyn, but that a few were charged to the club. He continued: 'There is no doubt the rules of the Club have not been strictly adhered to, and have been altered from time to time, as for instance, years ago it was optional for a quarryman to subscribe to the Club or not as he pleased, but for some years past membership has been compulsory, and looking upon the Club as intended for the benefit of its members, the Pensions, Donations, Relief and Deaths list, were, when the funds admitted of it, increased according to the exigencies of the cases, and as the members of the Club have come in for the benefit of it, I do not think I should be called upon to refund these items.'

W. J. Parry listed some of the discrepancies: in 1844 a disbursement of £5 did not appear at all in the year's summary; and the standard payment of £5 on death had been exceeded in the following instances:

1842	5 at £10, 1 at £7
1843	3 at £10
1844	1 at £7
1854	1 at £7

A number of coffins had been paid for by the club between 1842 and 1857, although Lord Penrhyn was 'always ready to pay for them'. In 1862–3 there were two separate entries for £3 after the death of John Jones; and Parry interpreted this as indicating a double payment. A number of unauthorised payments had been made for long periods during the years 1864–74; and an unauthorised pension of 2s 6d a week had been paid from 1855 to 1860. Lord Penrhyn, when informed of this state of affairs, paid over £2,000 and undertook not to make any prosecutions.[74]

In 1875 new rules were established for the Penrhyn Quarry Benefit Club; and these gave control of the club to the quarrymen. The general committee included the president, the vice-president, the treasurer, the secretary and two managers of the quarry. The elected members of the committee were two workmen from the Slate Mill and one quarryman from each gallery in the quarry. Lord Penrhyn was president, and Arthur Wyatt was vice-president. Every member was to pay a subscription of 1s every month on the quarry pay-day; and when any member was seriously hurt he was to be treated at the Quarry Hospital or visited every week at home by the surgeon. In case of illness a payment of 5s a week relief was to be made for twelve months; but this was not to be exceeded. A pension not exceeding 3s 6d per week was to be paid to any member who, because of old age, accident or lunacy, was unable to follow his work. When a member of 10 years standing died, his widow or next of kin was to be paid a sum of £5; and 10s was to be paid for every year's service under 10 years.[75]

In January 1902 the rules and regulations of the Benefit Club, now renamed the Penrhyn Quarry New Club, were published. All persons employed at the quarry were to be members, and the object of the club was to provide from members' contributions the following benefits: a sum of money on the death of a member, or the member's wife; relief during illness; the subscription of a sum not exceeding £100 to any nursing fund in the Bethesda district, to provide a visiting-nurse to the men and their families during sickness; and to subscribe not more than £13 per annum to St John's Ambulance Association. Lord Penrhyn was to be president, and E. S. Douglas-Pennant and E. A. Young were to be vice-presidents. The subscription was 1s 8d a month, and relief was 6s a week; but no relief was to be given if the illness was brought on by an 'irregular way of living or by immoral habits'. If any member was seen to frequent a public house whilst in receipt of relief, his relief pay was to be stopped.[76]

Many Friendly Societies were formed in the slate-producing districts; and these included the Temperance Societies, which were set up after the Temperance Revival of 1835. The Friendly Societies in North Wales were god-fearing and respectable, and they were favoured by the employers.

The influence of Calvinistic Methodism on the lives of the quarrymen was very powerful. The early leaders of the Methodist Revival were Howell Harris, Daniel Rowland, William Williams, Howell Davies and Peter Williams. Sunday Schools were established by Thomas Charles of Bala in 1784, and they attracted adults as well as children. Evan Richards or Richardson, a native of Cardiganshire, joined the Methodists and started a school in Brynengan, which became the centre of a revival in Eifionydd. Richardson eventually reached Caernarvon, where he opened another school. Education was almost as important to Methodists as their religion; and during the early nineteenth century the Methodist movement was responsible for the spread of schools along the coast from Bangor to Penmaenmawr.

Methodism gained its hold in North Wales with the outbreak

of religious enthusiasm displayed in the revival of 1817, which centred on Beddgelert. Methodist membership was greatly increased; and new leaders who appeared included John Jones, who was a quarryman at Talysarn and a shareholder in the Dorothea Slate Company. The Methodist revival was not an isolated phenomenon, however: between 1780 and 1830 the Baptists or Independents were also busy building chapels in Caernarvonshire. Wesleyan Methodism, with its doctrine of 'free grace', also came to North Wales at the beginning of the nineteenth century.[77]

The interest in the Chapel reduced interest in politics among the quarrymen; and instead of political divisions, there was a cleavage between Church and Chapel. The 1846 report of the Commission of Inquiry into the State of Education in Wales showed that the standard of teaching English to the working classes was low; and this, unfortunately, increased hostility towards Nonconformists and towards the Welsh language. Landlords, including the Pennants and the Assheton Smiths, contributed to the Church or National Schools; and in 1842 such schools were being attended by more than 100 children in Llanllechid, 60 children at Waunfawr, 100 at Llanrug, 65 at Vaynol, 26 at Capel Curig, 30 at Dolwyddelan, 46 at Beddgelert and 34 at Penmachno.[78]

British and Foreign Schools, which were undenominational and supported by the Nonconformists, were established in the 1840s in Caernarvon, Bangor and Penmachno. The land for these buildings was often provided by Anglican landlords, such as the Bulkeleys, the Pennants and the Assheton Smiths; and in certain circumstances these men also offered financial assistance. On 17 February 1858, for example, James Wyatt wrote the following letter to John Owen, the 'Secretary for Building the British and Foreign Schools' in Caernarvon:

> Colonel Pennant says that being a Church of England man he feels that his first duty is towards the schools connected with that Church, that although he has in some instances given sites where none could be obtained from other persons,

> yet that he has not been in the habit of subscribing to the British and Foreign Schools. He hears however from the Vicar of Caernarvon that the population of that place is sufficient to fill the schools already established and the one proposed. He has therefore requested me to forward to you £25 towards the erection of the building which I have now the pleasure of doing.[79]

A school for girls was established at Llandegai by the first Lady Penrhyn soon after her husband's death in 1808; and a school for boys was built there by the Pennants in 1884.[80] Favourable treatment of quarrymen who were members of the Anglican Church, although denied by the quarry owners, was a source of bitterness among quarrymen in both Penrhyn and Dinorwic Quarries by the end of the nineteenth century.[81] By that date religion and politics had become more closely associated: the Nonconformists as a rule had Liberal sympathies, whereas the Churchmen tended to remain Tories.

The slow growth of unionism among the quarrymen was in part a result of Methodism. The emphasis on respectability and godliness made the Chapel communities unwilling to drop their 'servitude' for the role of 'agitators'. Discipline in the quarries was another factor responsible for the small number of malcontents: heavy money fines were imposed for breaches of the rules, and quarry owners had an unwritten agreement not to take each other's men until they were satisfied they had left the quarry 'honourably'.[82] At Dinorwic Quarry in 1893 there were said to be no money fines; and no person was suspended for a first offence. Dismissal, however, followed a third offence.[83] Defiance of the manager's orders was penalised, at Penrhyn Quarry in 1892, by instant dismissal. For breaches of the rules, dismissal followed after the fifth offence.[84] At Penyrorsedd Slate Quarry in 1899 lateness was punished by the loss of half a day's pay.[85] Severe penalties were also imposed for violation of the safety regulations, since this endangered the lives of other men.

There can be no doubt that some of the quarrymen, under the paternal cares of the owners of Penrhyn and Dinorwic and of the larger Merioneth quarries, felt a genuine liking for their overlords. In January 1846, when Colonel Edward Douglas Pennant married Lady Louisa Fitzroy, the quarrymen shared fully in the festive occasion. 'Banners and flags' gave the town of Bangor 'a most animated appearance'; and the town guns fired salutes at regular intervals. Bonfires were lit, including a large one at Garth Point and one on the Bangor Mountain behind the Liverpool Arms Hotel. At Port Penrhyn shipping displayed flags, and there was an incessant discharge of artillery. At Llandegai tea and plum cakes were provided for the 230 pupils at the expense of Colonel Pennant; and at the quarry all work was suspended, the men receiving 2s each from Colonel Pennant to compensate for the suspension.[86]

Even Lord Penrhyn's son, who took charge of the quarry in 1885 and subsequently became extremely unpopular with most of his men, tried initially to foster a feeling of goodwill between management and men. In 1886 he defrayed one-third of the cost of a trip to London for the men; and on 15 October his agent E. A. Young reported that the men were 'extremely grateful and had thoroughly enjoyed themselves'. Officials at Euston told Young that they had never seen a 'more orderly set of Excursionists'.[87]

12
DEPRESSION AND DECLINE, 1882–1918

The years 1851–74 had been a period of expansion in the Welsh slate industry, and in business concerns generally. In 1850 the excise duty, which had been levied on bricks since 1784, was repealed; and brick production rose in the 1850s until it reached the boom of 1857. The number of slate-workers increased from 7,946 in 1861 to 13,576 in 1881. The boom which lasted from 1868 to 1873 saw prices generally rising higher than they were to be again until World War I. This boom was followed by a fall in commodity prices, which continued with fluctuations until the end of the nineteenth century.[1] The depression in trade reached the slate district a few years after 1873, and by 1878 the reduction in demand for slate had hit the quarries. The fact that during the prosperous period quarry owners had had a position of near monopoly had resulted in high prices being charged, sometimes for inferior slates. This had led to an attempt by builders to find substitutes for slates and thus break the monopoly. In the 1870s, however, importation of foreign slates was small; and tiles, though becoming increasingly popular in the South of England, were about 30 per cent higher in price than slates. In 1877 the selling price of best quality slates (size 24in × 14in per mille of 1,200) was £15 5s 0d; this compares with a selling-price of £7 17s 6d in 1851.[2] After 1877 prices were reduced; and this checked for a while the growth in demand for foreign slates.

Depression in the building trade was the main reason for the reduced demand for slates in the 1880s. Industrial disputes in the building industry and the slate industry were another reason for the contraction of output in the slate mines and quarries in

North Wales. The largest output ever achieved at Penrhyn Quarry was 130,000 tons in 1862. By 1882 this had fallen to 111,000 tons. The output at Dinorwic Quarry fell from 98,000 tons in 1862 to 87,000 tons in 1882. The quarries and mines of Nantlle, and Ffestiniog, however, developed during this period; and in Nantlle the output rose from 30,000 tons in 1862 to 74,000 tons in 1882. The Ffestiniog slate mines shipped 67,000 tons from Portmadoc in 1862 and 145,000 tons in 1882. This increase was due to the expansion of the export trade, the peak of which was reached in 1889.[3]

The stagnation in the building industry in the 1880s resulted in the number of buildings erected between 1878 and 1890 showing an increase of only 2 per cent, compared with 15 per cent between 1871 and 1878. The 1890s saw a recovery in the building trade, and in the slate industry; and the number of houses built in the UK increased by 14·8 per cent between 1891 and 1901. At this period suburbs were becoming increasingly fashionable, and they provided some of the demand for slates. House building averaged about 130,000 houses a year during the six years from 1900 to 1905; but from 1906 to 1909 it averaged only 90,000 a year, and in the years 1910–13 the figure dropped to 62,000.[4]

The records of the Llanberis Slate Company from 1878 to 1928 provide an insight into some of the difficulties facing a company with about 200 workers during the period of the slate industry's decline. In September 1879 the secretary, William Wotherspoon, reported 'another very serious reduction in the price of slates'; and as a result, the directors decided to put the men on short time. In April 1880 the new secretary, George Barrack, told the agent, H. Balfe, that he was not confined to one district, but was at liberty to make 'a journey North' if he thought it would be productive of favourable results.

The output of slates in 1880 was 4,861 tons; and in 1881 discounts varied from 10 to 25 per cent according to sizes and quantity. In February and March 1880 the company sent out letters, seeking orders for their slates from slate merchants in Swansea, Gloucester, Bristol, Cheltenham, and elsewhere.[5]

Another reduction in prices was made in January 1883; and in November 1883 a set of letters asking for orders was sent to merchants in Penrith, Llanelly, Oldbury and Southampton. The discount in 1883 was 15 per cent for credit or $2\frac{1}{2}$ per cent for cash; but by the end of the year the former figure had been reduced to $7\frac{1}{2}$ per cent for sizes from 16in × 10in downwards, as the demand for small slates was 'very brisk'.[6]

In December 1884 the managing director, John Menzies, organised a successful meeting of quarry proprietors and their representatives at Caernarvon. The quarry representatives who attended were the following: John Menzies (Llanberis Slate Company), the Hon W. W. Vivian (Dinorwic), Arthur Wyatt (Penrhyn), A. H. Dunlop (Oakeley), R. Greaves (Llechwedd), R. H. Phillips (Welsh Slate Company) and Norman Davies (Dorothea). It was unanimously agreed that no change should in future be made in prices without a meeting of the Caernarvonshire and Merionethshire quarry owners. Caernarvon prices were fixed about 5 per cent above those for Penrhyn and Dinorwic, with discounts of 10 per cent. In the case of Cefn Du, this meant that they would actually be selling at prices about 5 per cent below those of Penrhyn and Dinorwic Quarries. The objections to making any large advances in the prices were: the depression in trade generally; the proposal to increase the German tariff on slate; and the depression in the slate trade with America. Menzies expressed the hope that this meeting might have 'established something like stability in the slate trade'; and until his death in 1907 he acted as a go-between for all the Caernarvonshire quarries.

The lock-out at Dinorwic Quarry in October 1885 was described by Menzies as causing a 'good deal of sensation in the district'. He did not know what outcome to expect; but he considered that 'for the moment it ought to benefit the small quarries if the lock-out should continue for some time.' In February 1886 the secretary of the company wrote to slate merchants in Scotland where small sizes of slates, under 16in × 8in were used. Letters were sent to merchants in Dundee, Arbroath, Montrose, and Banff; and the merchants were advised that 'owing to the long stoppage of the Dinorwic Quarries and no immediate prospect of a settlement, it is probable that you may be inconvenienced for a supply of slates.' The 'Colonial Market' was said, at the time of these requests for orders, to be taking all the best slates in the sizes 20in × 10in and 18in × 10in. The lock-out at Dinorwic ended in March 1886. In May 1886 the secretary said that the company wished to deal direct with 'the consumers' in Glasgow, and not through a merchant. This was a practice never followed by the larger quarries, such as Penrhyn and Dinorwic.

In the 1880s speculative builders, particularly in the Midlands, were taking advantage of the low rate of interest and building houses with cheap slates. In September 1887 the secretary wrote to a firm asking for orders, as he had been informed that a 'deal of small second slates' were being used in the Midland counties, and they had a 'good and even lot' of these on hand. The terms were 25 per cent discount, or 2½ per cent for cash. The depression still continued in 1889; and in September 1889 Menzies declared that the slate trade was in an 'extremely depressed state' with stocks accumulating and prices falling; Lord Penrhyn had 'commenced this week to work 5 days a week'. It was, Menzies considered, 'almost impossible to work any Quarry to pay at present in the face of a fall of nearly 50 per cent in prices since 1877.'[7]

From 1880 to 1889 the domestic demand for slates declined; but foreign demand increased, and the Sales Ledger Cash Books of Penrhyn Quarry show the fluctuation of total sales as follows:

1886	£206,972
1887	£175,789
1888	£196,914
1889	£196,284 [8]

At Penrhyn Quarry export prices in the 1880s were lower for some blue slates than the home prices. There was a reduction of 11s on blue slates 18in × 10in, making an export price of £4 16s; blue slates 16in × 10in were £4 10s, a reduction of 10s;

and blue slates 16in × 8in were £3 6s, a reduction of 6s. These reductions were made in order to gain a foothold in the German market, where the blue-grey slates of Ffestiniog had preference. On 23 March 1886, however, it was said that the city of Melbourne had a 'large speculative demand and no prejudice against Blue or in favour of certain sizes only'. The output of slates from Penrhyn Quarry was reduced by the short time working which began on 3 May 1886, and by the prohibition of all overtime working. E. A. Young feared that this would weaken the quarry's hold on the German market, at the season when trade with Germany was usually good. He foresaw that the Germans would be forced to go to Caernarvon slate concerns. The reduction of prices to German customers was reported on 17 June 1886 to be worrying the Portmadoc export traders, as they were 'frightened' that Penrhyn Quarry's efforts to expand its German trade might succeed. Blue slates were in plentiful supply at Penrhyn but red slates were scarce. In June 1886 the Austrian empire, a principal customer for such slates, was about to impose a duty of £11 per ton; and so the merchants pressed E. A. Young to supply red slates as quickly as possible. Blue slates were harder to sell, and competition became fierce. On 12 July 1886 E. A Young complained to the manager of Oakeley Quarry that Portmadoc vessels were delivering from 1,270 to 1,280 slates per thousand, although 1,260 was the number agreed upon at the meetings of the quarry representatives. The Oakeley Company were allowing 3 per cent discount for cash, instead of the usual 2½ per cent; and their loading charge was only 1s per ton. In return for the Oakeley Company's guarantee that only 1,260 slates would be put into the vessels, E. A. Young undertook to confine his reductions in price to slates from 22in × 12in downwards, and to withdraw the reduction in respect of larger sizes after the close of the month.

Cut-throat competition and the depressed state of trade resulted in the closing down of some of the smaller quarries; Bryn Hafod y Wern, for example, closed in 1885. In November 1886 Lord Penrhyn was informed by E. A. Young that the Welsh Slate Company had decided to run their quarry while they had it, producing as much as possible and selling at any price that suited them. As a consequence of this policy, the London trade was reported to be 'dead'; but in 1888 the Welsh Slate Company was taken over by the Oakeley Company.[9]

Between 1889 and 1918 the quantity of slates exported fell from 79,912 tons to 1,592 tons. The decline in the German demand for Welsh slate, which in 1876 had represented 72 per cent of the total quantity exported, was the main reason for the falling off in the export trade. In 1910 the percentage of slates sent to Germany had fallen to 54.[10] This drop affected the Ffestiniog district most severely, as before 1913 Germany took from 35 per cent to 45 per cent of the slate shipped from Portmadoc: this amounted to about 39,000 tons in 1894, about 21,000 tons in 1900 and about 11,000 tons in 1910.[11]

The decline in the export trade was brought about largely as a result of the imposition of duties on imported slate which began in the 1890s. By 1910 slates imported into France and Switzerland were liable to duties of 8s 2d and 12s 2½d per ton respectively; and by the same year slates imported into Canada and the USA were liable to *ad valorem* duties of 25 per cent and 20 per cent respectively. There was no duty, however, on slates imported into Denmark; and no duty was charged in Victoria or New South Wales, though other Australian states levied *ad valorem* duties ranging from 5 per cent in Western Australia to 25 per cent in Queensland.[12]

The list of countries to which slates were consigned from the UK in 1913 was as follows:

	Tons
Denmark (including the Faroe Islands)	2,584
The Netherlands	481
Belgium	485
France	664
Germany	12,248
Other Foreign Countries	1,404
The Channel Islands	433
Australia	8,581

		Tons
New Zealand		324
Other British Possessions		875
	Total	28,079 [13]

The export of slates from North Wales declined from 72,715 tons in 1891 to 44,794 tons in 1898; but the home demand improved, and this period saw a return to prosperity in the slate industry. The reason for the revived demand for slates was the boom in the building trade. This was an industry closely linked with home investment, and the increased amount of building in the 1890s took place because British investment abroad was low. A programme of slum clearance schemes in towns such as Birmingham, Liverpool, London and Glasgow, and the erection of municipal buildings such as libraries, schools and public baths, meant that unemployment in the building trade fell from 4·1 per cent in 1894 to 1·8 per cent in 1896 and remained below 2 per cent until 1900. The demand for slates was so great in 1895 that the directors of the New Welsh Slate Company Ltd declared that there was 'nearly a famine in the slate trade' and that business was better than it had been for 20 years. Both Lord Penrhyn and Assheton Smith advanced the wages of their quarrymen by 5 per cent; and in several of the Ffestiniog concerns notices were posted announcing an advance of 3d a day in the wages of every class of workman.[14]

Quarries were reopened in response to the demand for slate; Bwlch y Slater Quarry, for example, was reopened in 1896.[15] Some of the larger quarries were paying good dividends as a result of the prosperity: the Votty and Bowydd Slate Quarries Company Ltd made a profit of £12,814 in the year 1895, and an interim dividend of 20 per cent was declared.[16] At Penrhyn Quarry E. A. Young reported to Lord Penrhyn in May 1892 that the selling price was 2s 6d per ton better than the previous year: in 1891 it had been £2 9s 5d per ton, and in 1892 it was £2 11s 11d per ton. The cost of production, on the other hand, had been 6d per ton lower than the previous year: £1 8s 3d per ton in 1891 and £1 7s 9d in 1892. Young added, 'I know you will find these very satisfactory, but please don't take too sanguine a view of the future.' Agitation over the Home Rule Bill in 1893 seriously affected the amount of building in Ireland, and exports to Ireland were consequently reduced.[17] The number of persons employed in the slate industry increased from 12,162 in 1891 to 16,766 in 1898. The output of slate in Wales increased from 352,186 tons in 1891 to the record output of 488,883 tons in 1898. The Penrhyn dispute of 1896–7, which lasted 11 months, created a scarcity of slates, as it reduced the output of Caernarvonshire from 252,911 tons in 1896 to 185,054 tons in 1897. In the following year, however, the output of Caernarvonshire was 318,012 tons, which was a record.

In May 1898 E. A. Young wrote to Lord Penrhyn:

> The Trade as a whole continues to take every slate as fast as we can make them, but there are indications in some places that Americans are doing our Market serious injury—for instance in Dublin American slates are selling at 25 per cent less than Penrhyn and we have shipped none to Dublin for over two months instead of the usual cargo a week. In Bristol again we are almost entirely driven out of the Market. However, the Australian demand has revived tremendously and therefore with the help of a good demand in the Midlands and London, all sizes are quickly sold.
>
> In the winter we shall probably get slack and begin to stock slates.[18]

Shipments from slate quarries in the USA had been exported to Britain from the 1870s, but it was not till 1898 that American slate affected the home market. In some circles there was a feeling that this foreign competition might not be a bad thing. In January 1896 the *Liverpool Daily Post* complained that the North Wales slate producers had formed a combination, which had 'enabled them to control the market, and practically place the whole of the building trade at their mercy.' Complaints were made by the builders of South Wales, not only about rising prices but also about difficulty in getting orders promptly executed. In 1896, however, the American quarries were regarded as too small to be able to compete with the quarries of North Wales and their 'slate lords'. In November 1895, however, E. A. Young was asking London slate merchants not to order any American slates and promising in return to keep them supplied with Penrhyn slates. In December, too, he wrote to W. Vivian of Dinorwic Quarry asking him to give an extra supply to the London firms and saying, 'I am sure it would practically choke them off and leave the Yankees to supply them to people we would rather not trust; this they are now doing in Ireland and also in South Wales district.'[19] The pressure on supplies of slate, at this period, was so great that the secretary of the Llanberis Slate Company wrote in March 1897 to a customer that they could not 'possibly allow any Discount, as we could sell our make ten times over at present terms and are even offered more by some.'[20] The principal American slate-producing states in 1897 were Pennsylvania, with exports worth £350,000, Vermont with £120,000, Maine with £25,000 and Virginia with £20,000; these were 'followed closely by the states of New York, Maryland and Georgia'.

In November 1897 R. B. Pritchard of the Eagle Slate Company, New York State, wrote to his relations in Caernarvon that conditions were very good in his quarry. He reported that a group of Londoners had been to see his quarry, and that he had sold some of his red slates and some of his green slates to them; they represented a company to 'foster the American slate trade in Great Britain and the Continent by the importation of slate from the USA'. Pritchard was well pleased at the thought, and considered that these were the best people to introduce red slates to 'prejudiced England'.[21]

Slates were also imported from France, Belgium, Norway, Portugal, Italy, the Netherlands and Germany. In France the largest production of slates came from the department of Maine-et-Loire and from the Ardennes; and Belgian slates came chiefly from Namur. The French exports of slate amounted to 46,000 tons in 1899, with imports not above 1,100 tons. Slate quarrying in the Ardennes had begun at least as early as 15 June 1551, when a municipal charter had been granted to the quarrymen.[22]

An aggressive attitude towards the import of American slate was taken by the Welsh slate companies. On 30 November 1898 E. A. Young sent letters informing the most important slate merchants, including Adlard, Braby, Etridge and Matthews, that, 'In order to enable Penrhyn Slates to hold their own in your district against the existing influx of American slates I consider it requisite on and after 1st January 1899 until further notice to invoice your usual ratio of Penrhyn supplies (by water only) in Best Blue and Best Green and Wrinkled 24in × 14in to 18in × 10in inclusive at List price without the 5 per cent premium, less 2½ per cent Cash discount. You will please clearly understand this special price refers only to slates for use in the London District.'[23]

No sudden decline in the industry set in after the peak year of 1898, when at Penrhyn a record profit of £133,000 was made.[24] In 1901, however, there was a depression partly as a result of the Boer War; and this was followed by a high bank rate and by scarcity of money. Unemployment in the building industry averaged 6½ per cent in the years 1901–5 and over 10 per cent in the years 1906–10. The state of the building industry affected the slate industry, and the number of persons employed fell from over 13,000 in 1909 to just over 8,000 in 1918. The output of slate fell from 402,184 tons in 1909 to 101,315 tons in 1918. The average value per ton fell from £2 16s in 1898 to £2 10s 7d in 1914, although from 1905 and especially from 1909, the prices of commodities generally had begun to rise again.[25]

The three-year dispute at Penrhyn Quarry led to a fall in output between 1901 and 1903. This meant that for a short time some quarries benefited from the scarcity of slate occasioned by the dispute; and although there was a partial return to work by some quarrymen in June 1901, quarries such as Dorothea were in the position of having a strong demand for slates, and were able to obtain the list price plus 5 per cent, and less 2½ per cent for cash. They could have obtained a higher price but the manager considered that it was better to make no change as they 'knew not the day the strike might come to an end', and did not want to disturb relations with their men. In fact they were faced with a short strike in February 1903 when the manager had closed an eating-house because of 'horse-play' on the part of the younger men. This strike was a minor one, but it annoyed the company since 'it got into the Press and the circumstances were magnified there.'[26]

The scarcity of slates produced by the dispute at Penrhyn stimulated the trade in Portuguese and French slates. Imports of American slates, on the other hand, were falling. The quantities were as follows:

Year	*Belgium*	*Portugal*	*USA*	*France*
	£	£	£	£
1898	13,252	9,549	267,479	39,908
1900	13,108	9,379	177,964	45,599
1902	12,521	23,788	128,054	105,892 [27]

Before the dispute Penrhyn Quarry had produced about one-quarter of the total output of Welsh slate, so it was not surprising that there was a deficiency. Even so, foreign slates formed at that period only a small proportion of the home consumption. With the complete resumption of work in Penrhyn Quarry, the importation of foreign slate fell from 119,805 tons in 1903 to 87,157 tons in 1905.[28] After 1903, however, there was a continuing general depression in the slate industry.

In October 1905 working days at Llechwedd Quarry were cut to four a week. Wages at Penrhyn Quarry were reduced by 10 per cent; and the number of workers was reduced by 150 at Penmachno and by 100 at the Co-operative Quarry of Pantdreiniog.[29]

The effect of the imported slates on trade was discussed in the *Master Builders' Association Journal* in February 1906. The countries exporting slate to the UK were then: Holland, Turkey, Italy, Canada, Norway, Newfoundland, Portugal, Belgium, the USA and France. Norwegian slates were discharged at London, Hull, Manchester and Kirkwall; Newfoundland had markets in

London and North Shields; Portugal supplied London, Bristol and Plymouth; the Belgian consignments went to London, Liverpool, Bristol, Grimsby, Hull, Lynn and Southampton; the USA affected sales in London, Liverpool, Bristol, Cardiff, Hull, Newport, Swansea, Dublin and Belfast; and France had sent large quantities to London, Boston, Bristol, Dover, Exeter, Faversham, Folkestone, Grimsby, Ipswich, Lynn, Newcastle-on-Tyne, Newhaven, Newport, Plymouth, Poole, Portsmouth, Rochester, Shoreham, Southampton, Teignmouth, Yarmouth, Dublin and Limerick. The Midlands, however, were less affected by foreign slates.[30]

In 1906 wages in the Ffestiniog district fell 10 per cent, and in some cases 15 per cent, below what they had been the previous year. 'Several hundred men' were dismissed from the Llechwedd Quarry in March 1906; but in June an improvement in the Nantlle slate trade was reported.[31] By 1906 however, another competitor had arrived on the scene. This was the tile, which was now in great demand because the cost of production had been reduced by the manufacture of tiles from asbestos fibre and cement. Slate was also being ousted from some of its traditional uses: earthenware had supplanted slate slabs for sanitary purposes, and the use of slate as partitions for slipper baths had declined as teak and marble were being used. In the Talyllyn district nearly all the quarries had ceased working by 1906, and the rate collector told Dolgellau Board of Guardians that he had never experienced so much difficulty in collecting the rates. In April 1907 Ffestiniog Town Council reported that there were over £2,000 arrears in rates. It was said that 'the people were utterly unable to make both ends meet owing to depression in the slate trade.'[32]

The Ffestiniog district was badly hit by the depression, the loss of exports and the competition from foreign slate. Portmadoc slates were almost identical in colour with French and American slates; and they were extensively sold in London and on the south coast, where French slates sold at prices 20–30 per cent below those of Welsh slates.[33] At Port Penrhyn and Port Dinorwic stocks were piled up at the quays in 1908; but the managers were said to be optimistic, as they believed the disadvantages of cheap foreign slates would soon be realised. The blue slates of France, it was said, soon turned 'dirty grey'; the red slates of Germany were 'soft as clay'; and the rough slates of America were unsuitable for the roofing of houses.[34]

Another blow to the slate industry was the discouragement given by the Board of Education, on grounds of hygiene, to the use of writing slates in schools. In September 1908, too, Penrhyn quarrymen protested to the authorities of the University College of North Wales, whose new building was being roofed with slates from Pembrokeshire. The quarrymen regretted this because of the 'extreme depression' in the local slate trade; and they pointed out that they had subscribed funds for the construction of the college. The slates had been obtained from the Presely Quarry, which produced olive and natural rustic slates; and the authorities defended their choice on 'artistic' grounds. The slates had in any case cost only £300.[35]

At this time Welsh quarry proprietors were facing criticism, in technical journals of the building trade and elsewhere, for not supplying heavier and rougher slates. The proprietors pointed out that quarries would produce more of these slates, classed as seconds or thirds, if they were an economic proposition. They did not, however, fetch the price of thin slates, although with their extra weight and uneven surface they cost as much, because of additional carriage and laying costs.[36]

Tennyson had praised the beautiful quarries of Penrhyn and Dinorwic, and in 'The Golden Year' he had celebrated Llanberis:

> . . . high above, I heard them blast
> The steep slate-quarry, and the echo flap,
> And buffet round the hills from bluff to bluff.

In 1890, however, William Morris strongly criticised the 'Thin Welsh Blue Slates', which in his eyes were comparable with corrugated iron and zinc.[37] In 1910 an article in the magazine *Beautiful Homes* referred to a new product called 'asbestos shingles'. These were made of asbestos fibre and Portland cement: there were two shades, grey and bright red, and the cost was the same as for a first-class slate roof.[38]

The Finance Act of 1910 had a damaging effect on the building trade, as the possibility of an additional tax of 4s in the pound on the enhanced value of land caused a reduction in the number of houses being built. By 1913 the effects of the depression were very severe in Nantlle, where people were suffering from want of food; and it was felt that it was high time that the slate roof should be defended, instead of being jeered at and derided for its ugliness.

Many slate quarries and mines were forced to close during this period. The three co-operative quarries of Tanybwlch, Moel Faban and Pantdreiniog closed in 1911; and between 1914 and 1918 the number of slate mines fell from 29 to 12, with the number of miners reduced from 2,231 to 946. In 1918 only 19 quarries were being worked, and the number of persons employed had fallen to 2,194.

The war affected the slate industry very badly, as building almost came to a standstill. In 1914 the Oakeley and Llechwedd quarries reduced their working week to three days; and the younger men at Oakeley were urged to 'join the colours'.

The loss of the German trade was serious for the Ffestiniog district; and in Caernarvonshire 1,100 quarrymen were unemployed. Temporary buildings, such as the Trawsfynydd Artillery Camp, were roofed with asbestos tiles; and when the clerk of the Merioneth Local Relief Committee protested to the War Office, he was told in December 1914 that 'an extended use of Welsh slates' was 'impracticable at the present time'.[39]

In October 1914 Cilgwyn Slate Quarry closed. In November Robin Duff, the new proprietor of Dinorwic Quarry, was killed in action; and in April 1915 Lord Penrhyn's second son, the Hon George Douglas-Pennant, was also killed.[40]

In 1917, in response to an appeal from the Ministry of Labour for National Councils in major industries, the North Wales Slate Quarries Association came into being; but the Ffestiniog District Slate Quarry Proprietors' Association, formed in 1892, continued as a separate body. Hitherto, Penrhyn and Dinorwic

Quarries had always remained apart from associations; but this time they joined in. The association included about 90 per cent of the slate manufacturers in Great Britain, the greater part being in North Wales. It agreed on a price list, but the members were not bound to adhere to this. The principal function of the organisation was to negotiate with labour organisations about wages and conditions of employment.[41]

In 1917 slate quarrying was officially declared a non-essential industry; and quarries such as Rhosydd, Craig Ddu, Rhiwbach and Diphwys were closed for the remainder of the war.

13
INDUSTRIAL UNREST

One of the reasons for the instability of the slate industry after 1881 was the worsening of relations between management and men in many quarries, and the reduction of output by frequent and often lengthy strikes and lock-outs. The success of the two strikes in 1874, at a period of prosperity, had increased the membership of the North Wales Quarrymen's Union, so that in 1877–8 it reached 8,190. This was about two-thirds of the total labour force; but almost three-quarters of the members came from the lodges set up at Penrhyn and Dinorwic. The report for 1877–8 gave the following membership figures:

Dinorwic	2,200
Caebraichycafn (Penrhyn)	3,450
Ffestiniog	378
Nantlle	750
Corris	180
Bryneglwys	80
Llanberis	290
Port Penrhyn	100
Port Dinorwic	48
Bethesda	120
Waunfawr	230
Alexandra	110
Rhostryfan	118
Rhos	80
Dolwyddelan	50
Rhyd-Ddu	6 [1]

A General Council of the union was held in May of each year, and delegates were sent from the lodges in proportion to the number of members. Small lodges with from 12 to 99 members sent one delegate each. The General Council of 7 members was elected annually, and met four times a year with the Executive Committee. Monthly subscriptions of 6d were paid by the men, with boys under 16 paying 3d. During the depression after 1876, the union encouraged quarrymen to emigrate by lending a sum of money averaging about £3 to each quarryman who emigrated. This policy was eventually given up in 1883, as it proved too heavy a drain on the union's resources.[2] Membership was affected by the hostility of the quarry proprietors and by the depressed state of the industry; and in 1883 there were only 3,535 members.

By this date, relations between the workmen and management in Dinorwic Quarry were extremely strained. In November 1878 the men suffered a reduction in pay of 12½ per cent. This was followed in December by another reduction of 7½ per cent, and short-time working of from 3 to 4 days a week was introduced. Two hundred men were dismissed before the end of January 1879; and another 100 were dismissed in March. Further reductions in wages were made in April, August and November 1879 and April 1880.

In 1880 the Employers' Liability Act was passed. This Act was unsatisfactory in its workings, because of the high costs involved in claiming compensation: only where a trade union assisted did an injured person have any chance of winning his case. There was nevertheless deep resentment in 1881 when George Assheton Smith, by threatening the withdrawal of pension rights and of support for the Benefit Society, forced his men to contract out of the Act. In 1885 the Dinorwic men were allowed to have the benefit of the Act; but this concession came too late to undo the ill-feeling the initial action had caused.[3] Another constant cause of friction at Dinorwic Quarry was favouritism: it was openly admitted, for example, that those workmen who were active in the Conservative Party were rewarded by a monthly bonus of £1.

Conditions at Dinorwic Quarry became rapidly worse. In July 1885 53 men were suspended from working, because 10 men had broken a local rule; and on 12 October the men held a mass meeting, in working time, at a spot a few miles away from the quarry. There they passed resolutions of no confidence in the manager, John Davies, and in the chief manager, the Hon W. W. Vivian. Vivian was connected by marriage with George Assheton Smith and the men contended that he lacked experience of the slate industry. A deputation was appointed to meet George Assheton Smith and explain the men's grievances; but without meeting the men Assheton Smith ordered notices to be posted all over the quarry. These notices said: 'In as much as a mass meeting was held during work hours on Monday afternoon 12 October, in defiance of an order made in July last, when you were cautioned that such a meeting if held during work hours, would not be tolerated.

'Notice is hereby given that your services will not be required after Saturday, the 31 October, and that all Barrack furniture, tools, velocipedes . . . belonging to you must be moved before 12 o'clock on Saturday the 31 October.'

On 27 October 1885, according to John Menzies of the Llanberis Slate Company, the men appeared to be taking the matter 'very lightly'; but Menzies knew that Assheton Smith was 'very resolute'.[4]

The lock-out lasted until 1 March 1886, when the men accepted the terms negotiated through John Robinson, a proprietor of Talysarn Quarry in Nantlle. The terms accepted were as follows:

1. No man will be debarred from employment owing to his being on the Committee, or for taking part in any agitation.

2. A man's religion and political principles will have nothing whatever to do with his obtaining employment (and Mr Vivian indignantly denies that they ever had.)

3. The quarries will be managed by able practical men.

4. Employment will be found for every man the works are capable of employing, and if any men will be kept out more will be taken on as the proposed improvements pro-

gress; (in connection with this it is right to state that from two to three hundred men will be affected by the alterations proposed) but this will not touch the quarrymen at all.

5. The new rules will be administered in a spirit of fairness and justice to the men, and not harshly.[5]

The terms sounded honourable, but in practice the men gained little. The union's funds lost about £10,000 from the dispute; some of the men's leaders were victimised; and the quarry rules were not always maintained.

At the Votty and Bowydd concern there was a strike of about 70 labourers in March 1885, but this quickly collapsed when the men were threatened with dismissal.[6] More serious discontent, however, existed at Penrhyn Quarry, partly because of a slump in trade. Reductions in wages and the steady dismissal of men created feelings of insecurity and anger. In 1878 3,500 man were employed, but by 1893 the number had fallen to 2,500. In 1885 George Sholto Gordon Douglas Pennant, who was to be created Baron Penrhyn on 31 March 1886, succeeded his father as master of Penrhyn Quarry. It was rumoured that the new master had a grudge against his quarrymen for the part they had played in his electoral defeats of 1868 and 1880. On the latter occasion he had been crushingly defeated by the Liberal Watkin Williams QC:

Watkin Williams	3,303
Douglas Pennant	2,206
Liberal majority	1,097

The result had come as a surprise; and the defeated candidate had bitterly referred to the men of Caernarvonshire as 'foremost in falsehood'.[7] This slur was naturally resented, but a later interpretation claimed that he was referring to the particular falsehood of broken election promises. One of Douglas Pennant's first unpopular actions as master of Penrhyn Quarry was his response to a meeting held at Bethesda in 1885 to discuss Disestablishment. The crowd that was kept out of the meeting burst the barricades, but there was no other violence. Douglas Pennant's solicitors tried to identify the Penrhyn workmen who had attended the meeting; but their efforts were unsuccessful. On 16 March 1885, therefore, Douglas Pennant sent a threatening letter to all his men, indicating that 'henceforth lawless conduct which brings disgrace on the name of a Penrhyn quarryman will be met by dismissal from his service.'

This letter was followed, on 4 May 1885, by a notice which terminated the Pennant Lloyd agreement of 1874. Douglas Pennant's reason for this action was that the stagnant state of the slate industry had resulted in low selling prices, so that it was no longer possible to maintain wages at the 1874 level. The minimum wage was therefore abolished, and the standard wage was reduced to 25s a week. It was announced that all complaints of a serious nature were to be made to the chief manager; Douglas Pennant 'declined altogether to sanction the interference of anybody (corporate or individual) between employer and employed in the working of the quarry'. The declared object of this policy was to ensure that the views of non-unionists were represented; but the real aim was to crush the power of the union. A letter of protest was signed by three unionists and three non-unionists, two of whom later claimed they had signed because they had misunderstood the English version; and for this, the three unionists were dismissed in May 1885. They were reinstated in the following June, but the strong impression remained that unionists were marked men.[8]

In 1886 Arthur Wyatt was succeeded as chief manager at Penrhyn Quarry by E. A. Young, who was to hold that position until his death in 1910. Young was a keen businessman; and handled all Lord Penrhyn's private investments (many of which were in railway stock). During his period of management he came in for much adverse criticism from the men.

E. A. Young quickly set about putting the quarry in order. The head manager was discovered to be favouring certain workmen at the expense of others; and, with the help of the chief clerk of the quarry, various cases of corruption were discovered and reported to Lord Penrhyn. A report of 25 February 1886, for example, referred to 'Bad Rockmen . . . being paid £10 a month. And one case where two Partners in a Bargain have during the whole of 1884 and 1885 been paid a regular average of £9 7s per month.'[9]

Under Young's management the conditions of labour at Penrhyn Quarry continued to deteriorate. On 10 January 1888 a notice was put up at the quarry, terminating an old monthly holiday that had always been allowed to the men. In September 1890 one of the men's annual holidays was taken away from them; and despite protests it was not restored. By the beginning of 1891 discontent was so widespread that a committee of workmen was elected; and on 10 July 1891 a deputation met the manager, asking for an increase of wages to the following daily rates:

Quarrymen and rublers	5s
Workmen or Bad rockmen	4s 6d
Slate Loaders	4s 6d
Labourers	4s
Miners	5s
Sawmen in the mill	4s 9d
Smiths and moulders	5s

The men requested more liberty in the selection of partners; monthly lettings for the 'rybelwrs'; greater security of contract so that when men had a poor bargain for two months, the letting of the third month would enable them to make up the deficit; and restoration of the last Wednesday holiday at the end of the monthly letting, on the understanding that all workmen would 'square up their slates and take their bargains that day.'[10] Young refused these requests, but offered to close the quarry 'as before on pay Saturdays'. This, he told Lord Penrhyn on 28 July 1891, was 'in order to split them and turn the day men etc against the quarrymen'. At the same time Young also wrote to the London slate merchants, Roberts, Adlard & Company, telling them that the men were very dissatisfied and would appeal to Lord Penrhyn but that they would still get the same reply. Young continued:

> but besides this all the other Quarrymen are doing their best to get the Penrhyn men out to reap the benefits themselves

and they *may* succeed. Some of the Caernarvon proprietors are also acting in such a way as to back up the men in their attitude.

Lord Penrhyn gave his reply to the deputation on 20 August 1891, saying that he 'saw no reason for varying the reply given by Mr. Young.' He went on to explain to the men why the 25 per cent wage increase they were demanding could not be given:

> The slate trade is now suffering very severely owing to the many labour agitations and strikes that have taken place during the last few years, in Australia, on the Continent and in many parts of England, and the idea held by many that a strike would cause prices to rise and be followed by a rise in wages is absurd; prices of slate might advance temporarily at the smaller Quarries where the production is limited, but the more the price is rising—cheaper material will be used.
>
> There is another very clear reason why your wages cannot now be advanced and it is a fact which all of you know very well, viz, this—the rock in many parts of the Quarry is not now turning out as favourably as it used to do and the poundage and tonnages are higher, and the cost of production has consequently considerably increased . . .
>
> I may also take the opportunity of repeating what I have before now told you, viz, that when the state of the Slate trade shows such an improvement as to justify the step, I shall have much pleasure in giving a proportionate rise in wages.

Despite Lord Penrhyn's words, trade improved in 1890 and 1891, and profits rose from £45,000 in 1890 to £55,000 in 1891. While E. A. Young was away giving evidence to the Royal Commission on Labour in February 1892, the men held meetings 'throughout the Quarry'. Young wrote to Roberts, Adlard & Company on 15 February 1892 that these meetings would probably force him to take action sooner than he intended, 'and lead to a rumpus very shortly indeed'. On 25 April 1892 Young wrote to Sir John Ruleston about the plan of Sir John Gorst, member of the government, to address the North Wales Quarrymen's Union in Caernarvon. Young complained that this would give a 'fresh fillip' to the union, which was 'now practically discredited among the men'. An address by W. J. Parry or 'any other demagogue' would have been of no importance, Young argued; but for 'a high Government Official to take as it were men into his confidence, men who are utterly unworthy of trust, as unionists, is a step which seems to me most unfortunate and decidedly opposed to the interests of Lord Penrhyn.' The quarry's profits for 1892 were £89,871, and in April 1893 wages were increased by 5 per cent.[11]

Penrhyn Quarry was not alone in experiencing labour disputes. At Llechwedd Quarry in May 1894, one of the men was suspended for not returning to work at the request of the sub-agent. The following day, having been refused permission to hold a meeting, the men appointed three men to act as a deputation. When the owner, J. E. Greaves, ordered them to return to work 'one by one and through the office', they refused and came out on strike. The strike ended in October; but 'three or four members of the Committee were never allowed to work there again.'[12]

In the years 1891–5 membership of the union declined from 5,970 to 1,423; but in 1896 it increased, because of the Penrhyn dispute, to 6,540. The new trouble arose out of the men's desire to attend the Labour Day demonstration to be held at Ffestiniog on 4 May 1896. On 13 April E. A. Young was informed of this desire by a deputation. He refused to accept the deputation, and said that the men must apply individually. Another deputation then informed Young that the men proposed to attend the meeting at Ffestiniog in a body.[13] On 30 April Young spent some time in the quarry; but no one asked him for leave, and he described the committee's 'tactics' as follows:

> Each gallery has got a so-called representative, and he has been since to his Overlooker or Mr. Pritchard [one of the managers] with a similar message that all the men in his gallery would be away from work. They have . . . got a large

> number of men to sign in a book that they would not work on Monday, and of course the dodge is to prevent us knowing who go to Ffestiniog and who stop at home. However, it does not appear that more than half the 3,000 will be absent and therefore we have still decided to throw the Quarry open and thus throw the loss of wages on the backs of the Agitators.[14]

On 1 May 1896 E. A. Young wrote to Lord Penrhyn: 'The men's tactics on this occasion are very peculiar and I have come to the conclusion that at the instigation of headquarters the Penrhyn men are ordered to try and "pick a quarrel." And their ultimate view will be to re-establish the Union and the old Quarry Committee, which is really at work at the bottom of the present movement.'

E. A. Young was not optimistic about the outcome. He felt that after Labour Day some new demand would be made: 'then must come a "pitched battle" and apparently Penrhyn Quarry has been definitely selected for the trial of strength.'

The men who absented themselves from the quarry on Labour Day were suspended for two days, not for attending the meeting but (as the editor of the *Mercury* was told on 6 May 1896) 'for being absent without leave'.[15] On 2 July 1896 the men asked for a standard daily wage of 5s 6d for quarrymen, miners and sawyers, a minimum wage of 4s 6d, and a poundage of 10s for apprentices. This was refused; and on 7 August a statement of the men's complaints was sent to Lord Penrhyn. They condemned the system of contracts recently introduced into the quarry, because these contracts brought in middlemen who sometimes let contracts to inexperienced quarrymen. They also complained of the lowness and uncertainty of their wages. On 17 August 1896 a deputation of the men met Lord Penrhyn, who refused all their requests.[16]

Feelings at this time were running very high; and on 15 September 1896 E. A. Young told W. Vivian of Dinorwic Quarry that matters were coming to a head: 'You may rest assured that I shall stick fast to discipline and retain the management of the Quarry in my hands come what may.' The struggle had now developed, as far as the management was concerned, into a battle between Capital and Labour.

Two men, Robert Owen and David Davies, were suspended on 15 September for having measured bargains with a tape-measure. They contended, however, that they had merely 'stepped' these bargains in order to provide the committee with accurate information. On 26 September, after failing to appear when summoned to the office, they were dismissed. A meeting of the General Committee was held to consider these dismissals; and this meeting decided not to continue with negotiations but to strike in March of the next year. On 28 September E. A. Young suspended 71 men—the members of the committee, together with those who had signed the memorial of 7 August—for 'misrepresentation, insubordination and inciting the men to strike.'

On 29 September the men refused to take their bargains unless they were given an explanation of this action;[17] and on 30 September the men were locked out. This lock-out, which Young called a 'strike', lasted until August 1897. The men tried to have the dispute settled by the Board of Trade under the Conciliation Act of 1896; but on 21 October Lord Penrhyn told Colonel Ritchie of the Board of Trade that 'no good is likely to result from any outside interference in the management of my Quarries.' Since the quarrymen had ceased 'by their own act' to be in his employment, he argued, there was no one for whom the suggested conciliation could act.

Some men, such as fitters, joiners and some of the daymen, continued to work in the quarry; and a 'dastardly attempt' to wreck the workmen's train was made by those locked out. The engine driver 'noticed that the line was blocked with large stones in a manner which would inevitably have wrecked the train and possibly have caused fatal injuries to the workmen had not the obstacle been providentially seen.' E. A. Young was the focus of much resentment; and on 7 April 1897 he wrote to Colonel Ruck, the Chief Constable of Caernarvonshire, that affairs at Bethesda were becoming 'more riotous'. He referred to a mass meeting on 'Saturday last'; and he reported that when his name was mentioned 'there were shouts of "shoot him" or "kill him" emanating no doubt from irresponsible youth, but nevertheless having a mischievous effect.' Young, who was a keen cyclist, then related how he had cycled through Bethesda to Lake Ogwen and back by way of Conway, a distance of about 50 miles. On his way, he had noticed that some of the men near Bethesda were 'very sullen and as I passed the new works at Llandudno Junction, a group of men, no doubt late Penrhyn Quarrymen thought fit to shout as I passed.' As a postscript, Young added the following statement: 'on absolute reliable authority . . . an ambuscade had been prepared for me over a wall and amongst some trees close to Ogwen Bank, with a view of stoning me as I came back on my bicycle from Ogwen Lake, but of course their design failed owing to my returning quite innocently in a different direction altogether. Besides this, your Inspector at Bethesda has been informed of the damage that has recently been done to workmen's cupboards in their Quarry huts etc, and as recently (presumably on Saturday night after the mass meeting) a hut near the Slab mill in which the foreman of the railway platelayers keeps his tools etc was broken into, all the windows smashed and everything inside the hut including the stove broken.'

On 13 May 1897 Young gave the Bangor Inspector of Police further information about the molestation of those who continued to work in the quarry. The cases mentioned included that of a carpenter, who was 'rather roughly treated in going from his work'. Young asked that these men might have 'proper police protection to enable them to walk about the streets without being continually insulted.'

After two meetings with Lord Penrhyn in February and March, terms of settlement were finally accepted by management and men in August 1897. The terms of the settlement were as follows:

I a) The grievance of any employee, crew, or class should be submitted by him or them in the first instance to the local manager. If the decision of the local manager was not accepted, then the grievance should be submitted to the

Chief Manager either personally or by deputation. The deputation should consist of not more than 5 employees selected from the same class as the person or persons aggrieved, who had to be included in the deputation.

b) Grievances in which the employees generally were interested, or which they might adopt on behalf of an employee, crew, or class who had submitted their grievances under the preceding clause and were dissatisfied, could again be submitted to the Chief Manager by a deputation consisting of not more than six employees.

c) In all cases of importance an appeal might be made to Lord Penrhyn, either by the individual or by a deputation, against the decision of the Chief Manager; the grounds of such appeal should in all cases be first submitted to his lordship in writing.

II Suitable Rybelwyr would be given Monthly Bargains without delay as soon as the Management found it practicable.

III The letting of contracts was to be left in the hands of the Management who were to engage all persons employed thereon and see that each employee received his just ratio of wage.

IV Previous to the cessation of work the average wage paid to the Quarryman was 5s 6d per day, other piece work classes being in proportion, (viz. Badrockmen 4s 7d and Labourers 3s 7d); when work was resumed this same basis was to be continued so long as trade permitted.

V All the late employees who wished to work in the Penrhyn Quarry would be readmitted, in a body as far as practicable, and the remainder as soon as work could be arranged for them. Reasonable time would be allowed to those who might then be employed at a distance.

Congratulations on the settlement of the dispute were received from John Menzies of the Llanberis Slate Company, and from other quarry owners. To John Menzies, E. A. Young wrote on 23 August 1897 that the men could have had the same terms 'months ago if they had not been so badly advised'. To J. G. Ashmore, manager of the Oakeley Quarry, he wrote: 'many thanks for your congratulations. I have no doubt as you say the result of such a beating will be a lesson to them to be content in the future when they are well off, but, unfortunately they never seem to realise when they are well off.'

At the end of August 1897, E. A. Young reported to Lord Penrhyn that the number of men who had applied to be taken back in the quarry was about 150 short of the number employed before the dispute. This, he concluded, was 'fairly good proof of how glad they are to get back again'. David Davies and Robert Owen, the two men who had been dismissed for 'stepping' a bargain, and for insubordination in not answering a summons to the office, were reinstated on 2 September, having expressed regret 'for the great mistake' they had made.[18]

The right to form a combination had not been conceded in the terms of settlement, and despite assurances given to the men that there would be no 'black list' of men being refused work as a punishment for their part in the dispute, about twenty-five were not given employment, on the grounds that because of drunkenness, incompetence or unfitness for work through physical handicap they could not be re-employed. The men's deputation were not satisfied with this reply, but they were unable to take the matter further.

During the time the men were locked out, £19,161 19s 7d was subscribed for them, mainly through the efforts of the *Daily Chronicle*, and by means of concerts and general subscriptions.[19] This public support was encouraging, but after the return to work most of the workmen were in a state of anxiety and insecurity. E. A. Young, however, wrote to W. Vivian on 29 November 1897: 'The men here are now working very satisfactorily and seem to be much happier even than they were before.'[20]

The economic background to the unrest in the 1890s, in contrast to the previous decade, was one of prosperity for the proprietors, since slates were in great demand during the building boom. By 1898, however, the peak of the boom was reached, competition from foreign slates increased, and a slow decline began in the slate industry. The number of unionists fell to 1,158 in 1899, and this was largely from a fear of victimisation. On 8 July 1898, E. A. Young wrote to David Pritchard, the manager: 'I have marked 51 boys that you may take in at once, after carefully looking through them again to see that none of them are sons of any of those who took any active part during the strike.' In December 1898 W. R. Evans, the chairman of the committee, was dismissed, because he had disagreed with David Pritchard who had at once reported the matter to Young. Evans pleaded to Young to be taken on again, asking for his '50 years' loyal service' to be taken into consideration. Despite these pleas, Young refused to reinstate him, adding a private postscript to Lord Penrhyn, as follows: 'W. R. Evans [was] about 62 to 65, of rather poor physique and no apparent sign of intellectual power, in fact not in any way a man I should have expected to see others elect for their chairman, but of course my view of him was purely superficial.'[21] The following June, Robert Davies, the chairman of the deputation which had met Lord Penrhyn several times, was also dismissed. Net profit of the quarry for the year 1898 was reported in January 1899 by Young to be £133,000.[22]

In April 1900 orders were given to the Penrhyn workers that union payments were not to be collected in the quarry, as had formerly been allowed. The reason given for this prohibition was that some of the men had complained, and Young told Lord Penrhyn on 4 May that he was 'sure many were being bullied into paying'. On 26 October there was an outbreak of violence in the quarry against certain contractors in Ffridd Gallery. Young gave his account of the incident to Lord Penrhyn:

> A short time ago I had some trouble with 14 men in this Gallery, all badrockmen, and the ring leader was suspended for 6 days for insulting the overlooker . . . one crew was also suspended for 3 days . . . [I decided] to move 6 men to various other parts of the quarry last Tuesday. The same day they came and said they would rather work elsewhere than in

> the Quarry. The following Wednesday Edward Williams, a Contractor, was sent up there with four or five others. He was warned not to go to Ffridd Gallery, but he ignored this and started work. Shortly afterwards some of the young men in the gallery above began stoning the Contractor's men and shouting 'Hurrah'. Edward Williams had to run for it. He reached the *Red Lion* having come down the slate incline and sought shelter in the weighing machine office. There was a crowd of from 300 to 500, and Williams was struck by a piece of slate. He was then dragged and driven down to the rubbish tips towards the slab mill.

Eventually the men went back to work; and Williams, seriously injured, was sent home in a cab.

Young continued his report by saying that an attempt had been made to identify the rioters and prosecute a few of the offenders. 'Of course', said Young, 'the whole point is this, that these bad-rockmen of whom some may possibly have worked for several years in that gallery are bringing forward that old claim to the right to work at their own bargain, year after year, without being disturbed. This claim I know you are quite determined not to allow, in fact it would be quite impossible to manage the quarry without having a free hand to shift the men at any time it is found necessary. If however they do get up an agitation with a view of making it general throughout the Quarry which I do not in the least anticipate, the ground of their agitation will not be simply the removal of these men from one gallery to another, but against the system of contracts which they so stupidly still persist in calling the sweating system. I have cautioned Mr. Mears to be very particular when inflicting punishment for any breach of Quarry Rules etc during the next few weeks so as to avoid as far as we can irritating any of the men, and for the present I have told him on no account to send anyone to work at any of the bargains in question nor is he to take the slightest notice of the fact that several of the officials, who not only became involved in the crowd, but who spoke to the men asking them to leave Edward Williams alone, . . . declare it impossible for them to identify any one of the rioters . . . I understand the rioters have very little sympathy from the general body of the Quarrymen and a delay in taking any visible action will tend to allow matters to cool down until a more sensible view is taken. I have purposely refrained from going personally to the Quarry and have no intention of going up there at present, in fact . . . it is my intention, if you do not disapprove, to go away on Tuesday for a week's golf at Great Yarmouth . . . Please therefore do not on any account alter your plans for the sake of coming here at present.'[23]

Young was sanguine about the 'riot' and its outcome, but in eight days' time there was an outbreak against another contractor and his two sons. Lord Penrhyn then instituted proceedings against 26 men, and before the men were tried, the management dismissed all of them. Penrhyn workmen went down in a body to Bangor on the day of the trial, and in consequence they were suspended from work for fourteen days. The cases at the first hearing were adjourned, and at the second, the workmen again walked to Bangor in procession. Only 6 of the men tried were convicted and fined; and in the meantime military forces had been called in by Colonel A. Ruck, although this action was condemned by various public bodies, including Caernarvonshire County Council. The men went back to work on 19 November, but eight galleries were not let. By 21 November the men were suspicious, and the next day they went to the quarry but did not start work. Some time in the morning E. A. Young telephoned to tell the men to 'go on working or leave the quarry quietly'. The men then walked out.[24] Thus began a dispute which lasted three years.

On 23 November 1900 the men entrusted their case to the Executive Committee of the union; and on March 1901 the General Federation of Trades Unions made a report on the situation. It was felt that up to the point when the men resumed work after the suspension, they had acted wrongfully; but it was also said that if 'reasonable care' had been taken after the court proceedings the lock-out would have been averted. The grievances of the men were listed as follows:

1. The Union was not recognised.

2. The management, while partially recognising collective action, did all in their power to discourage it, even to the extent of discharging those who went on deputations.

3. The men desired the reinstatement of certain victimised men.

4. The men were prevented from holding meetings in the quarry, although the men's homes were scattered over a very large area.

5. A minimum wage was not recognised. The nature of the rock varied largely, so much so that on bad rock poundage was allowed to make up their wage; the granting of poundage was in the hands of the 'letters'. In letting work, or setting bargains, they tried to fix the price so that the men could earn 27s 6d per week. The men desired a minimum of 4s 4d per day, but the management claimed that this would mean having to discharge over 200 old men who could not earn that amount.

6. The men desired the abolition of the contract system, but were willing to take contracts co-operatively.

7. The men objected to the bullying of the contractors and subordinate officials.

8. The rules of discipline were harsh, the most serious probably being the fines, by which the men who might be 15 minutes, or less, late, lost a half-day's pay, over 15 minutes a whole day; the fines went to the sick-club.

9. The men desired a more democratic management of the Benefit Club.

10. The men wanted permission to have an annual holiday.

These grievances had been submitted to E. A. Young at a conference held between him and four representatives of the men on 19 December 1900. The result was that Young would not agree to make any alteration with respect to nos 1, 2, 3, 4 and 5; but in regard to no 6 he agreed to try a co-operative system as an experiment in a part of the quarry where no contract system had before existed. Young promised with reference to no 7 to deal with authenticated cases brought to him. The fines for being late were reduced from one day's to half a day's pay. He promised to put the Benefit Club under a more democratic management; and no 10 was postponed for further consideration. The men took a vote on whether to accept Young's proposals or not; and by 1,707 votes they rejected them. Young then offered to reduce the fines for lateness to suspension of a quarter of a day, but matters still remained at a deadlock. The report of the Federation of Trades Unions expressed fears that unless something was done to bring the two parties together the dispute would last until the men were literally starved into submission.[25] This proved to be right.

Many of the men were suffering great hardship. Only those who had 12 months' membership of the union received any union support, and many had to depend on support from the Relief Committee, which included W. J. Parry. During the early part of 1901 pressure and threats were brought to bear on the men to persuade them to return to work on E. A. Young's terms. Paragraphs were sent to the press reporting that Lord Penrhyn was about to open the quarry; and there were rumours that as many as 1,500 had sent their names in to E. A. Young accepting his terms. An announcement was finally made that the quarry was to be opened on 11 June 1901, and a large number of extra police were brought into the neighbourhood. On opening-day, a gold sovereign was given to each of the men who returned; the number is uncertain, but was probably about 400. Those who returned, according to W. J. Parry, included some of the men who had caused trouble in the previous October and November. Others were said to be men who had been formerly discharged from the quarry, for various offences such as repeated drunkenness or poaching on the Penrhyn estate. The gift of a sovereign was said by Lord Penrhyn to be a spontaneous gesture on his part, a response to the sight of the men's pinched faces; it was not in any way intended as a bribe.[26]

Bitter animosities and distrust abounded in Bethesda after the reopening of the quarry. On 20 July 1901, the returned workmen sent a resolution to Lord Penrhyn, thanking him for his kindness in giving them a rise of 5 per cent and a bonus of 5 per cent at the end of the month. They acknowledged Lord Penrhyn's 'constant kindness' towards them, especially in 'times of sickness and distress'. On 27 July, however, the Home Office received a petition signed by 600 persons, together with a statement by the Chief Constable that the police force was not adequate to preserve the peace. A detachment of the Queen's Bays was therefore sent to Bethesda. In August the public houses at Bethesda were closed at 9 pm by order of the Bangor magistrates; and at Bangor Police Court, a number of quarrymen were fined for 'obstruction in the streets, for intimidation, and assault'. In September the detachment of the Queen's Bays gave a military tournament in Penrhyn Castle Park. Some of them were invited by Lord Penrhyn to ride on his private railway to the quarry, and the party was shown round the grounds of Penrhyn Castle. They departed from Bethesda for Aldershot soon afterwards, having become 'very popular with the more peaceably inclined of the inhabitants; and their departure was witnessed by a host of friends.'[27]

The year 1900 had witnessed a number of strikes and lock-outs, in addition to the Penrhyn lock-out. The list included the Thames lightermen, the Taff Valley Railway employees and the Staffordshire pottery workers. At Dinorwic Quarry, however, the management of the quarries was said to be materially improved under W. Vivian; many 'objectionable features' had been swept away. On 1 January 1902 Vivian retired, and he was succeeded by Ernest Neele, 'late district manager of the Chester & Holyhead branch of the London & North Western Railway.'[28]

Bethesda continued in a very disturbed state, particularly on Saturday nights; stones were thrown, windows were broken, and there were a number of assaults. On New Year's Eve 1901 there were many scenes of violence, including attacks on private houses and on two public houses which had supplied drink to the 'traitors'. On being informed of this, Colonel Ruck appealed for military aid; and a company of Staffordshire Infantry and half a squadron of the 7th Hussars quickly reached the district. The presence of troops further angered some of the strikers; and as the police escorted some of the quarrymen home, stones and abuse were hurled at them.[29]

In October 1902 a sub-committee of the Standing Joint (police) Committee heard evidence about the state of affairs at Bethesda, and the prosecutions undertaken by the police. Evidence was produced to show that the people of Bethesda had been asked, shortly after the reopening of the quarry in June 1901, to display a notice in their houses saying 'No traitor in this house'; and it was said that the return to the quarry by some of the men had been the 'sole cause' of ill-feeling. Most of those who had returned to the quarry lived at Tregarth; and although there were public houses there, 'hundreds' of these men went to Bangor every Saturday night. Many chapel-goers had been forced by intimidation to go to the church; and the attitude to the police had changed from the old harmonious one to one of hostility.[30]

On 29 November 1902 the men made an application to the Board of Trade to have their dispute settled by arbitration. The Board of Trade, after considering the matter, replied in December that action by them would be unlikely to achieve a speedy settlement. Other attempts at conciliation, including approaches to A. J. Balfour, Lord Roseberry and the Caernarvonshire County Council, were similarly unsuccessful.[31]

The *Daily News* and the *London Daily Chronicle* sent correspondents down to Bethesda to report on the happenings. *What I saw at Bethesda*, a pamphlet written by the *Daily News* correspondent in 1902, declared that in Bethesda High Street there was 'a stream of men and women on their way to the railway station.' Some were leaving with their families, and others were leaving their wives and children behind. 'As the train steams out, the women try bravely to raise a cheer, a cheer that ends in a sob.' The correspondent described the ruin that had overtaken Bethesda: 'The shops have now few customers. Fourteen hundred of the men are away.' Many were in the well-paid coal-pits in South Wales; others were earning a pittance in the Nantlle quarries. 'Misery and privation are present in hundreds of the quarrymen's cottages.' He described bare but clean homes with starving children inside; the families were dependent on grants from the Relief Committee.[32]

Some of the men found employment at the co-operative quarries of Pantdreiniog, Moel Faban and Tanybwlch, which were registered in July 1903 as the North Wales Quarries Society Ltd. W. J. Parry was the manager of Pantdreiniog, which opened in August 1903 with 230 men. The experiment did not prove a success; and many of the workers became disillusioned about the benefits of co-operation, as relations between the men and management became strained over familiar topics such as low wages and dismissals. The co-operative quarries were all closed by 1911.[33]

In March 1903 Lord Penrhyn won a libel action against W. J. Parry for statements which had appeared on 22 June and 13 July 1901 in the *Clarion.* The first of these statements, which were deemed to be libellous of Lord Penrhyn 'both personally and as an employer of labour', was as follows:

> We must not lower our banner now. It must be a fight to the death with this tyrant. He will stoop to anything to gain his end. He has brought down the traitor's price from 30 pieces of silver to 20. He gave each of the men who had returned to work and those who had continued at work during the lock-out a gold sovereign. Who but Lord Penrhyn would have done this? His friends say he is a good sportsman, but evidently he can strike below the belt. I do hope the papers and unions will come out strong and continue their support.

Of this statement, Parry said that it was not intended for publication, but that, as it contained his 'honest expression of Lord Penrhyn as an employer of labour, he could not withdraw and apologise.'

The second statement was as follows:

> Lord Penrhyn can congratulate himself on the rise of the death rate here already, and there is no need for thinning

> them with the military unless he is dissatisfied with the rise.

The damages and costs of the case against Parry amounted to £2,508; and these were partly raised by public appeal.[34]

The funds raised in support of the strikers reached £41,358; but by the end of the year 1903 the men's spirits were broken. At the weekly meeting in Bethesda on 14 November 1903, a vote was taken in favour of ending the struggle; and that evening the information was sent to Lord Penrhyn.[35] In his reply, Young said that applications for work could be made to him by letter, and that he would inform the applicants 'as soon as possible if and when they may resume work at the quarry'.[36]

The cost of the Penrhyn dispute was immeasurable. The financial loss, in wages, was reckoned at about £364,000; but in terms of suffering and bitterness, never completely effaced, the damage was still greater.

The depressed condition of the slate trade in the years after the dispute resulted in wages being reduced; and men were re-employed only a few at a time. Lord Penrhyn died in March 1907, and was succeeded by his son. E. A. Young died in October 1910, and was succeeded by W. Hobson. On 11 September 1911 a loyal address from the Penrhyn quarrymen thanked Lord Penrhyn for 'his kindness in granting us twice a 5 per cent bonus on our wages on the occasion of the coronation of the king and the investiture of the Prince of Wales. We also express a hope that the good feelings now existing between us will continue believing that the conditions of success are in co-operation between master and workmen.'[37]

Membership of the union fell to 1,250 in 1905; there were very few members in Penrhyn, and only about 200 in Dinorwic. Conditions at Penrhyn Quarry were not improved until 1916, when a system of training apprentices was introduced. In 1918 the men obtained weekly payments, and a minimum wage of 4s 6d a day. At the same date the union, which had been reorganised by R. T. Jones, was recognised by the owners of Penrhyn, Dinorwic and Oakeley. Many proprietors in the Llanberis,

Nantlle and Ffestiniog areas had accepted the union from an early date.

The Penrhyn dispute caused a shortage of slates, which encouraged merchants to import foreign slates and use cement and asbestos tiles. The quarrymen gained nothing at all by their heroic stand against oppressive conditions. Lord Penrhyn and E. A. Young, who were both members of the Free Labour Protection Association established in London in 1897, believed that they were making a stand against the threat of socialism. W. J. Parry was not a Socialist, however, but a Liberal; and there was little justification for Lord Penrhyn's fears that his quarry was being taken from him by his employees.[38]

14
THE CHANGING STRUCTURE OF THE INDUSTRY, 1918–1945

At the end of World War I there was an urgent demand for houses; and in England and Wales, between March 1919 and March 1933, the number of new houses built was 2,062,000. Unemployment figures in the building industry fell from 16·8 per cent in 1922 to 9·8 per cent in 1925. The General Strike in 1926 sent the percentage of unemployed builders up to over 13; and in 1936, when there was a serious trade depression, the percentage reached 28·9 per cent.[1] The fluctuating situation in the building industry affected the slate industry. In the immediate post-war period, when there was a great demand for slates, output increased and many companies became profitable again. The employment and output figures were as follows:

End of Year	*Number Employed*	*Total Output (in tons)*
1919	6,604	164,098
1920	8,304	215,269
1921	9,520	237,350
1922	9,523	231,410 [2]

Prices of slates advanced rapidly; and by 1 July 1920 the prices per mille were well over three times the pre-war figure. (See table on page 285).

A minimum wage was agreed on by quarry owners and union officials in October 1918. It was not welcomed by all proprietors; but T. Williams, a Dinorwic Quarry official, wrote to W. Vivian:

Year	*24in × 14in*	*24in × 12in*	*22in × 12in*	*22in × 11in*
	£ s d	£ s d	£ s d	£ s d
1913–16	15 15 0	14 5 0	12 0 0	11 0 0
1917	17 2 6	15 5 0	12 2 6	11 0 0
1918	19 12 6	17 10 0	13 17 6	12 12 6
1919	27 0 0	24 0 0	19 0 0	17 7 6
1920	52 15 0	47 0 0	37 2 0	34 0 0

Year	*20in × 10in*	*18in × 10in*	*18in × 9in*	*16in × 10in*
	£ s d	£ s d	£ s d	£ s d
1913–16	8 17 6	7 0 0	6 5 0	5 12 6
1917	9 17 6	7 10 0	6 10 0	6 2 6
1918	11 5 0	8 12 6	7 10 0	7 0 0
1919	15 7 6	11 17 6	9 17 6	9 2 6
1920	30 0 0	23 2 6	19 7 6	17 15 0

Year	*16in × 8in*	*14in × 8in*
	£ s d	£ s d
1913–16	4 5 0	3 2 6
1917	4 15 0	3 15 0
1918	5 10 0	4 5 0
1919	7 2 6	5 10 0
1920	14 2 6	10 15 0 [3]

'I hope you will agree, seeing the unsettled state of labour and the trend of the times, that the settlement made was the best could be done in the circumstances.' The rate was 5s 9d per day for quarrymen, 5s 6d for rock labourers, and 5s 6d for labourers. In 1919 the minimum rate for quarrymen was 8s 3d; and a scale of wages was also obtained for apprentices. In 1926 the North Wales Quarrymen's Union was affiliated to the Transport and General Worker's Union; and in that year a sliding scale of wages was put into operation. The scale varied with the selling price of slates, the ratio being fixed at 20s a ton in selling values and 1s 2d a day in wages; this meant that for every £1 reduction in the average selling price of slates there was to be a reduction in wages of 1s 2d per day.[4] The wages system in Penrhyn Quarry was improved after 1919, as the workers on the rock face and the workers in the slate mill became independent of each other. The rockman was given a piece rate for every ton of slate blocks he extracted, and a lower rate for every ton of rubbish he removed. The blocks were divided among the quarrymen in the mill, who had a uniform list of prices for converting blocks into slates.

Between July 1920 and July 1922 wages fell from 12s 6d per day to 9s 4d; but from 1922 to 1930 wages fluctuated very little, because the selling price remained almost unchanged.[5] In 1930, however, the selling price of slate was reduced; and in 1931 prices were about 9 per cent below the 1922 level. Slates 24in × 14in cost £39 12s 6d per mille, and slates 14in × 8in cost £9 7s 6d per mille. Wages in 1931 were £2 9s a week for day labourers and £2 2s a week for piece workers; and the standard of letting bargains was £2 14s 6d. In 1932 the quarry proprietors of Nantlle and Ffestiniog asked their workmen to accept a reduction of 10 per cent in their wages, but the men refused. The proprietors of Dorothea Quarry then asked the men to accept a 5 per cent reduction, but this also was refused. Sir Richard Redmayne, a former chief inspector of mines and quarries, was asked to settle the dispute; and he recommended a substantial reduction in wages.

The new scale of wages, which was in force by November 1932, was as follows:

	Letting Standard (per day)	*Day Rate* (per day)	*Minimum* (per day)
Quarrymen	9s 1d	8s 1d	7s 3d
Rock labourers	8s 7d	7s 7d	6s 10d
Labourers	8s 4d	7s 4d	6s 8d

Apprentices	
1st year	1s 8d per day
2nd year	2s 5d " "
3rd year	3s 3d " "
4th year	4s 1d " "
5th year	5s 0d " "
6th year	6s 6d " " [6]

Working hours were reduced from $47\frac{1}{2}$ per week during the summer months in 1919 to 46 per week during the summer months in 1933. Despite this reduction in hours, the quarrymen were only slightly better off in real wages than they were in 1914; and employers complained that the reduction of hours had decreased output per head. In 1913 the output of finished slate per man per annum was 33 tons for mines and 32 tons for quarries. On 29 January 1919 W. Hobson, the manager of Penrhyn Quarry, gave his opinion of reduced hours: 'Personally I am very keen upon doing all that can be done to prevent the shortening of the working hours in the Slate Quarries. I think the average actual working hours with us are about 7½ hours, of course after deducting the meal hour. And then there is a lot of time when the men are not working, such as blasting time, and if you add to this the loss of time such as holidays, attending funerals, hay harvest, rough weather, the actual working time becomes small.'[7]

The problem of falling output caused more labour-saving devices to be introduced; these included excavators and cranes. Output per man at Dorothea Quarry rose slightly between 1933 and 1938, as follows:

Output Per Man Per Month in Slate Made

1933	1·47 tons
1934	1·50 ,,
1935	1·53 ,,
1936	1·58 ,,
1937	1·75 ,,
1938	1·74 ,, [8]

The average daily earnings of workmen at Dorothea Quarry in 1941 were 12s 8d for quarrymen, 11s 4¼d for rockmen and 10s 2½d for labourers. By that date, therefore, they had returned to the level of 1920.[9]

The break-up in 1933 of the North Wales Slate Quarry Owners' Association, which had helped to keep slate prices stable, led to increasing competition among the quarries. As a result, there was a considerable fall in the value of slates. In October 1934, for example, Votty and Bowydd were selling 'Best

Old Vein 20in × 10in' at £22 4s 6d for 1,200, with a 2½ per cent rebate.[10]

The trend in the slate industry towards amalgamation became more pronounced after World War I. In 1918 the Amalgamated Slate Association was formed to take over the Alexandra, Cilgwyn and Moeltryfan Crown Quarries, with an authorised capital of £50,000. During the prosperous years 1919 and 1920, the company paid a dividend; and in 1922 the number of men employed was 415. In 1931, as a result of the trade depression, its quarries were closed; but thanks to a government loan of £13,000 they were reopened in 1932.[11] In 1921 the Dorothea Slate Company Ltd purchased the South Dorothea Slate Quarry at a price of over £7,000. The company had already taken over Penybryn Quarry in 1894; and this quarry was worked until 1932, when about seventy-six men were dismissed. In 1933 Galltyfedw Quarry was worked by the company.[12] The Oakeley Company, which by 1900 had acquired the Holland and Rhiwbryfdir Companies, the Welsh Slate Company, and the Cwmorthin Quarry amalgamated with the Votty and Bowydd Slate Quarries Company Ltd in 1933. Diphwys Mine, which had closed in 1927, was reopened by the Oakeley Company in 1936. The Maen-offeren Company acquired Rhiwbach Quarry in 1928; and in 1930 the Park and Croesor Quarries and the Rhosydd Quarry closed.[13] In 1936 the Ffestiniog slate industry was hit by an industrial dispute, and thus lost many working days. The dispute was settled by arbitration.

Costs of production in the slate industry continued to rise during the years 1939–45. Most of the accessible and productive rock had already been extracted, so that quarries became deeper and the expense of transporting slate to the mills increased; and the proportion of waste rock to finished was also increasing. In 1932 Caernarvonshire Employment Exchange Committee produced an analysis of the costs of slate production. Wages, at 75–80 per cent, were the main item of expenditure. The other costs were: salaries, 2 per cent; health, unemployment and workers' compensation, 4 per cent; stores and freight, 3 per cent; bank charges and general expenses 1 per cent; coal and power 6 per cent; depreciation of machinery 1 per cent; and repairs to machinery and spare parts, 2 per cent. At Llechwedd Quarry the total costs of production for four weeks ending 16 February 1935 were £5,194 for 730 tons of slates.[14] As wages were not high, economies could be achieved only by increased mechanisation, which involved a large investment of capital. Transport charges also played a large part in the increasing costs. By 1921 loading expenses had risen from 1s to 4s a ton, and railway costs from 12s 6d to 25s 6d a ton. Before 1914, slates were largely shipped by water in small sloops, barques and other vessels; but as a result of the war, most of the sloops disappeared, and rates for transport by water rose from 6s and 7s a ton to 40s and 45s a ton. The bulk of the slates were therefore sent by rail, which was expensive and often involved breakages. After 1921 many proprietors sold direct to builders or slaters; but the slate merchants, who bought from quarries and distributed to consumers, continued to exist.[15]

Very few quarries were able to make a profit in the years just before World War I, but dividends were again paid after 1918 in some concerns. The Dorothea Company was one of the paying concerns; in February 1919, a final dividend of £10 a share was paid out of the profits of the previous year. An interim dividend of £15 per share was paid in August 1919; the profits for the half-year ending in June had been £4,885. In February 1920 a final dividend of £45 per share was paid; and in February 1921 a final dividend of £3 15s per share was paid. The quarry profit in July 1921 was £13,054, and an interim dividend of £3 per share was declared. A dividend of £3 15s per share was paid in 1922; but in the following year, no dividend was paid, although a profit of £10,107 had been made on the working of Dorothea Quarry. In January 1924, after heavy rain, there was a serious landslide which covered all the best bargains; but in 1925 a dividend of £3 per share was paid out. After 1926 dividends grew smaller; and in 1930 the interim dividend was 12s 6d per share. At this date, because of the depression in trade, the quarry was working only three days a week. In the years 1931–9 the dividend was 10s per share.[16]

The Votty and Bowydd Slate Quarries Company was less

fortunate than the Dorothea Slate Quarry Company. During the war years they made losses; and although a profit of £569 was reported in February 1921, no dividend was then paid. 'Abnormal expenditure' had been required to restore the quarry to its pre-war state; and new labour-saving machinery was installed, 'in view of the high wages and short working now in force'. The slate mills were electrified, 'the greater part of the cost of this being met by the sale of obsolescent or derelict steam plant'. The following year did not fulfil expectations, as 'like other quarries' the Votty and Bowydd Quarries were stopped for several weeks, 'owing to the unprecedented drought in the summer, which caused a failure in and a cessation of the Hydro-electric power supply by the local Power Company.' The drought was succeeded by several weeks of heavy rain, which culminated in a cloudburst; and after the collapse of an important catchment drain in the mountains, the mines were flooded. This disaster cost the firm thousands of pounds. In February 1928 the company paid a 10 per cent dividend on capital issued before 30 June 1927; but in 1931 and 1932 they suffered from the general depression of these years. The losses were £865 in 1931 and £568 in 1932; and in the latter year there was a 10 per cent reduction in the number of men employed. The 'unremunerative prices of slate' were blamed as the chief cause of the losses.[17]

Before World War I the decline in slate production was less marked than the decline in the building industry; but in the years 1922–38, when the building industry expanded, there was no comparable expansion in slate production. Between 1922 and 1938 the average annual output of slates in Great Britain was only 281,000 tons; whereas in 1913, when the number of new houses being built was small, it had been 353,000 tons. After 1924 the demands of the building industry were met by roofing tiles; and where slates were used, builders often chose the cheaper foreign slates in preference to the Welsh ones. In the 1930s there was a higher rate of building than in the 1920s; but the production of slate still did not rise correspondingly. The figures were as follows:

Years	*Average Number of Houses Built Per Annum in Great Britain*	*Average Tonnage of Slates Produced Per Annum in Great Britain*
1922–29	163,650	286,002
1930–39	274,691	278,239

The production of tiles expanded rapidly: in 1935 it was estimated at 1,200,000 tons, which was six times that of 1912. In 1935 the output of slates from North Wales was 271,000 tons, as against 364,000 tons in 1912.[18]

One of the factors which produced this startling result was the cheapness of mass-produced tiles; and during the 1930s local authorities were in search of the cheapest building materials they could find. In July 1933, for example, the secretary of the Oakeley Company wrote to several slate companies, including the Votty and Bowydd Quarries, about the LCC Housing Authorities' scheme for building 1,896 houses in one area during the next 21 months. Of these houses 1,467 were to be roofed with tiles and 429 with slate; and the LCC claimed that the cheap bricks being used, which were dark red, were 'aesthetically unsuitable to mix with grey slate'. New slum legislation for the eastern parts of London produced plans for tenements up to ten storeys high, most of which were flat roofed or roofed with tiles. The secretary of the Oakeley Company concluded that in the bigger towns slum clearance schemes would not create a 'very sustained demand for slates'.[19]

Suburban building continued during the inter-war period; and with wider streets and more detached and semi-detached houses the appearance of roofs became more important than it had been in the nineteenth century with narrow streets and rows of terrace housing. The resultant concern for colour benefited not only the cheap tiles but also the expensive slates of Cornwall, South Wales and the Lake District. In 1882 North Wales had contributed 85 per cent of the United Kingdom's total slate production; but by 1933 this had declined to 75 per cent.

Flat, asphalted roofs became the rule for many new buildings; and by the 1930s corrugated iron on sheds and farm buildings was being replaced by asbestos-cement sheeting.[20] The Oakeley Company attempted to overcome the competition of colourful tiles by producing colloidal slates. In 1936 they installed a plant at Blaenau Ffestiniog, where slates were covered with permanent tints. This plan, however, proved unsuccessful, as builders considered the colloidal slates too expensive; and in 1939 the attempt was given up.[21]

At a 1921 meeting of the Manchester Geological and Mining Society W. Hobson, the manager of Penrhyn Quarry, explained why the slate quarries were at a disadvantage when competing with the artificial tile trade. The tile manufacturer usually had his works in a populous area. When trade was slack, therefore, he could shut down part of his plant; and the superfluous labour could be absorbed in other industries in the neighbourhood. The slate quarries of North Wales, on the other hand, were in thinly populated areas where there was no floating labour population. They had to train their own workers; and if a man was discharged in slack times, there was no alternative occupation. Some of the men went to South Wales; but, said Mr Hobson, 'that is a foreign country as far as the North Walian is concerned.' Hobson gave the following analysis for the typical slate of North Wales:

	Per Cent
Silica	55·30
Oxide of iron	10·00
Alumina	24·84
Lime	0·36
Magnesia	2·46
Carbonic acid	Nil
Sulphuric acid	0·21
Potash	1·47
Soda	0·53
Water of hydration	4·70
Loss	0·13

The absence of carbon dioxide and the small percentage of lime and magnesia were proof, said Hobson, that perishable carbonates were absent; and this explained why most Welsh slates lasted many years, even in the smoky acid-laden atmosphere of large towns. Slates from some American quarries, however, were capable of lasting only twenty years; and some German slates were 'even worse', since they contained 5·2 per cent lime, 3·6 per cent magnesia, and 4·45 per cent carbon dioxide.[22]

The chemical disadvantages of foreign slates did not prevent French, Belgian, Norwegian, Italian and German slates from competing effectively against Welsh slates in the inter-war period; but American slates ceased to be serious competitors. In September 1934 the Federated Quarry Owners of Great Britain declared in a memorandum, which was circulated to Welsh quarries, that 'the increased importation of foreign roofing slates . . . constitutes dumping, which should be terminated by the imposition of duties at the rate asked for in the original submission in 1932, viz. £2 per ton.' In 1932 an *ad valorem* duty of 15 per cent had been levied upon imported slates; but the quantity imported had continued to increase. The chief reason for the importation of these foreign slates was their cheapness; in quality they were usually inferior to the Welsh slates, and after exposure to the atmosphere of British cities many of them became discoloured and decomposed. The best French slates were an exception, being much more durable. At Southampton and London in 1934 the following prices were charged for Welsh and Portuguese slates per mille of 1,260:

Southampton	*Portuguese*	*Best Welsh*	*Medium Welsh*
size 20in × 10in	£12 10s	£21 5s 6d	£20 4s 6d
London			
size 20in × 10in	£10 8s	£20 12s 5d	£18 17s 2d

The low prices of Portuguese slates were rendered possible by the low wages paid to quarrymen in Portugal; the rate in Portugal in 1931 was approximately 1s 6d to 3s 3d per day, as against the British rate of 8s to 10s per day.

The average price for French slates imported during the quarter ending March 1934 was £5 12s 0d per ton, as against the average price for Portuguese slates of £4 1s per ton. The greatest quantity of slates, chiefly from France and Belgium, was imported into Harwich; but they were also imported through Plymouth, Portsmouth and Grimsby. London was the main port for Portuguese slates; and Norwegian slates were imported through Leith, Grangemouth, Newcastle, North Shields and Harwich. Competition was keenest in districts far from the Welsh slate areas, as railway rates were relatively high compared with rates for transport by water.[23]

In 1889 17·5 per cent of British slates had been exported; but in 1932 the figure had fallen to 5 per cent. In the years 1925–9 the average quantity of slate exported from the United Kingdom was 10,736 tons. By the years 1935–8 this had fallen to 3,672 tons. Exports to the Irish Free State fell from 6,963 tons in 1923 to 5,256 tons in 1932. The export trade was seriously affected by the high tariffs on imported slates imposed after 1918 by Germany, Australia, and the Irish Free State.[24]

The letters received by J. W. Greaves Ltd of the Llechwedd Slate Quarry throw some light on the obstacles to the sale of Welsh slates on the continent. On 24 October 1927, for example, L. Hammerich and Co of Aarhus wrote as follows:

> We do not intend to import French slates, but other firms do, and are offering them at such very cheap prices as to make people reluctant to pay the higher prices for Portmadoc Slates. As a matter of fact French slates which we have seen here recently look exceptionally fine and very much like Portmadoc slates. There is no doubt that they will find a market here which will further decrease the sales of Portmadoc slates, which already has gone back much, unless you find it possible to lower your prices to meet this competition.

On 13 February 1928, the same firm wrote:

> We regret to report that several of our customers have discontinued entirely buying Portmadoc slates and have

> turned instead to French slates. We have recently seen a carload of French slates which looked excellent: fine blue colour, even thickness and not thicker or heavier than Portmadoc.
>
> Prices: 24in × 14in £17 4s + 10 per cent
> 24in × 12in £14 0s + 10 per cent
>
> Free on rail, French station, and even if the railway freight from there to Denmark is about double the ship's freight from Portmadoc you will easily see the enormous price difference.

On 22 February 1930 the same firm complained that Christensen and Nielsen of Aarhus had advertised a cargo of Portmadoc slates due in Aarhus in February; the Hammerich Company had 'not a single slate in stock', and their rivals would therefore be able to supply some of their customers.[25]

In 1933 the British Standards Institution decided in the interest of clarity to dispense with such conventional terms as 'Ladies' and 'Duchesses', defining sizes instead by length and breadth.[26]

Many attempts were made by proprietors of quarries to utilise the large amount of waste material produced in the process of manufacturing finished roofing slates. In August 1895 E. A. Young had written: 'Lord Penrhyn has lost so much money re Slate Dust experiments that I don't think him likely to go into any more—I had a scheme before me only last week for converting it into Portland cement, but it was worthless. The fact is all the schemes fail, so far, on account of the expense of pulverising the slate waste. We sent 1,350,000 tons of rubbish over the tips last year, and I expect to clear 1½ million this year.'[27]

In July 1919 the North Wales Development Company Ltd was floated with an authorised capital of £150,000 to produce slate dust products, such as paper and tyres, at Pantdreiniog Quarry in Bethesda. W. J. Parry was one of the representatives of the syndicate; and he approached several quarries, including Dinorwic, to obtain their slate waste. The company proved a disastrous failure, and was wound up in 1921. In the USA experiments were carried out on the utilisation of slate waste in cement making and on the manufacture of roofing slates from slate granules. In 1920, about 300,000 tons of slate granules were used for manufacturing purposes.[28]

In the 1920s large crushing mills were erected at Penrhyn Quarry, where the slate was ground to a fine powder and mixed with bitumen. This mixture was called 'Penrhyn Grout', and was used as a binding material in road construction. Slate powder known as 'Fullersite' was also produced: this was used for vulcanite and moulded rubber goods, for asphalt, for bituminous products, and as a base for paints. The manufacture of concrete building blocks from slate waste was also undertaken at Penrhyn Quarry in the 1940s.

Distribution of Slate Production in Great Britain in 1927 and 1937 Respectively

District	*1927* (Tons)	*1937* (Tons)
Caernarvonshire	170,744	165,779
Merionethshire	76,357	55,932
Denbighshire	4,559	2,975
Montgomeryshire	613	799
Pembrokeshire	617	439
Carmarthenshire	919	846
Cardiganshire	125	—
WALES	253,934	226,770
Lake District	12,546	18,716
Cornwall–Devon–Somerset	15,961	19,218
ENGLAND	28,507	37,934
ISLE OF MAN	2,383	8,052
SCOTLAND	13,447	12,695
GREAT BRITAIN	298,271	285,451 [29]

The outbreak of war in 1939 was followed by a great decline in the demand for slates, as new building was strictly controlled. By

1940 about 4,600 men had been compelled by unemployment to leave the quarries; and in the course of the war about 2,000 quarrymen joined the forces. The bombing of towns and cities after 1940 created an urgent demand for roofing slates; but the shortage of labour resulted in a neglect of development work, and reserves of readily accessible slate were used up. The number of apprentices declined from 735 in 1939 to 183 in 1946; and some small concerns closed down. In Merionethshire these included Diphwys, Wrysgan, Manod, Craig Ddu, Bugail and Rhiwbach; and in Caernarvonshire they included Gallt-y-Llan, Garmon Vale, W. R. Morris, T. Owen and Vronlog Green Slate Quarries Ltd.[30] Some of these quarries, however, later reopened.

15
THE SLATE INDUSTRY SINCE 1945

In 1939 the number of men employed in the slate industry in North Wales was 7,589. By the end of World War II this number had fallen to 3,520; and in 1972 it was under 1,000. As a result of war-time circumstances, only 18 out of the 41 concerns operating in 1939 were still functioning in 1945. In that year the number of concerns fell to 16; and the total output of roofing and damp-proof course slates was 79,248 tons. Penrhyn Quarry and Dinorwic Quarry employed more men and produced more slate than all the others put together. In 1946 Penrhyn Quarry employed 1,065 men and produced 20,745 tons; and in the same year Dinorwic Quarry employed 1,360 men and produced 30,975 tons. These quarries were still privately owned, Penrhyn by Lord Penrhyn and Dinorwic by Sir Duff Assheton Smith. About ten of the smaller concerns were also privately owned, and the rest were limited liability companies.

The North Wales Slate Quarries Association was revived in 1943; and by 1945 it represented the whole of the North Wales slate industry. Most workers employed in the slate quarries were members of the Transport and General Workers' Union. A Welsh Sectional Industrial Council for the Slate Quarries Industry was established after the war; it consisted of four members appointed by the North Wales Slate Quarries Association, and four members of the union. Its main function was to settle disputes in any section of the industry. In theory, wage rates were the same in all quarries; but in practice they varied, as they were paid sometimes on an individual basis and sometimes on a group basis. A group might include both rockmen in the quarry and splitters and dressers in the shed; but sometimes these two sections constituted separate groups. A price or 'let' was fixed for the work; and on this the wage was assessed according to the output of the individual or group, with a guaranteed minimum wage.[1]

In 1947 the following new wage-rates came into operation:

Day men	Letting standard	16s 2d
	Day rate	13s 11d
	Minimum rate	13s 3d
Rockmen	Letting standard	15s 7d
	Day rate	13s 5d
	Minimum rate	12s 8d
Labourers	Letting standard	15s 3d
	Day rate	13s 1d
	Minimum rate	12 4d

A five-day week was introduced in 1947; and hours at most quarries were from 7 am till 5 pm. In 1972 the hours at Llechwedd Quarry were from 7.20 am to 4 pm; and a 40-hour week was worked. Wages there began at £17.50; and there was a scheme of incentive bonuses, so that wages could reach £40 per week. At Penrhyn Quarry a 40-hour week was also worked; and a similar bonus-scheme operated. The average wage was £29 per week, and the old system of bargains had disappeared.[2]

In 1946, in order that the entire output might be used to repair war-damaged houses, the Ministry of Works prohibited the use of slates 13in and more in length for the roofing of new houses; and in the same year they refused to allow the prices of slates to be increased as much as the proprietors wished.[3] The number of men employed in the North Wales slate industry rose in 1946 to 4,050; but there remained a serious shortage of skilled labour, especially of rockmen and miners. Consequently, development could not always be undertaken; and although orders were received by companies for the popular sizes of slates, they were often not able to meet the demand. The reasons for the shortage were varied. During World War II some of the workmen had joined the forces, while other had been directed to war-work outside the slate districts. This resulted in a loss of population, as many never returned to their former work. The danger and difficulty of most slate production, too, discouraged quarrymen's sons from following their fathers' occupation; and parents were often reluctant to allow this.

The average age of the quarrymen tended therefore to become higher. Wages had always been low, even in the years of prosperity and expansion; and because of the emphasis on education in most slate districts many of the younger men proceeded not to the quarry but to professional jobs. Another serious handicap to the recruitment of labour was the widespread fear of silicosis.

In 1951 the number of men employed in the slate industry of North Wales was 3,059. The basic monthly wage at Penrhyn Quarry was £22 5s od. A daily bonus was also paid; and there were piece-work rates for the production of slate rock or slates above a certain amount.[4]

Labour relations were often unsettled in the post-war period. In 1949 Dorothea Quarry experienced a labour dispute which caused a partial stoppage of work for some weeks during the summer. The prospect of nationalisation, which was being suggested in 1949, undermined the confidence of some proprietors; and at Dorothea Quarry capital expenditure was suspended in view of this possibility and because of the 'impending general election'.

By 1952, however, Dorothea Quarry had replaced its obsolete steam pumps; and it was hoped that the new equipment would effect a considerable saving in fuel costs. In the same year the directors reported that sales were being affected by 'keen competition from other roofing materials'; and they also declared that the company's difficulties were being increased by 'ceaseless demands for higher wages'.[5]

The diary of Capt S. T. A. Livingstone-Learmonth, who was a salesman for J. W. Greaves's Llechwedd Quarry during the 1950s, reported various problems, including the competition from other roofing materials. In 1957, firms in Fishguard considered trade 'very, very quiet, in all directions'; and no foreign slates were being imported. In Bristol much building was being carried on; but no slates were being used, as most of the buildings were prefabricated offices and shops with flat asphalt roofs or tiles. In Plymouth people were said not to want foreign slates at all, as their experience of French ones had been so bad. Travelling through Cornwall, one did not see 'a single new, or even recent building slated'. The price of slates was so much higher than that of asbestos, clay or Marley tiles that even local councils in traditional slate districts would not have them. The Delabole slate quarry was 'said to be feeling the pinch very badly'. At Bournemouth some slating was being carried out; but this was confined to 'what one might call the carriage trade', which included churches and 'architect jobs'. Southampton reported a 'few foreign slates coming in', mainly from Portugal; but people were said to mistrust them. The rebuilding of Portsmouth had been carried out largely with tiles. It was said that the concrete tiles had 'knocked out the clay tiles completely' in the Portsmouth area, and that they were being used for 90 per cent of all roofing work. Most of the firms visited emphasised their preference of lorries to rail transport, since lorries were able to deliver to the door of the works.[6]

In the years 1952–5 imports of foreign roofing-slates were declining, as the following figures demonstrate:

	1952 Tons	*1953* Tons	*1954* Tons	*1955* Tons
Irish Republic	61	—	56	15
France	1,874	731	923	1,562
Portugal	1,718	2,086	1,947	2,004
Italy	1,596	362	260	428
Total	5,188	3,538	3,695	3,994 [7]

Exports of slates from the United Kingdom were also falling, as is shown by the following figures:

Slates and slate slabs	*1952* Tons	*1953* Tons	*1954* Tons	*1955* Tons
Channel Islands	224	141	145	34
Other Commonwealth Countries	263	119	90	123
Irish Republic	1,607	1,155	1,861	1,056
Denmark	176	162	141	159
Netherlands	138	346	325	418
Other Foreign Countries	39	28	107	8
Total	2,447	1,951	2,669	1,798 [8]

Imports of roofing slates for the years 1959 and 1960 were as follows:

	1959 Tons	*1960* Tons
Norway	440	526
France	249	471
Portugal	1,693	934
Other Foreign Countries	15	6
Total	2,397	1,937 [9]

The production of slate in Wales and Monmouthshire declined steadily from 1958 to 1970, as can be seen from the following approximate figures on page 303.

During the 1950s and the 1960s many quarries invested in modern machinery in an effort to increase output. In 1953 diesel locomotives were introduced into the Votty and Bowydd Quarries, where they operated the main haulage to and from the mills.[11] Dorothea Quarry invested heavily in modern machinery in 1964; and roads were provided so that heavy rock-working machines could reach the bottom of the quarry. In 1966, however, the effect of the credit squeeze began to be felt; and the Dorothea Company had to halt its expansion programme and concentrate all efforts on finding new markets at home and abroad. A separate department was set up to produce slate for flooring. In addition, a slate craft section was formed to supply the gift trade; and the firm sent exhibits to the International Gift Fair at Blackpool.[12]

Year	*Tons*
1958	54,000
1959	46,000
1960	52,000
1961	45,000
1962	40,000
1963	33,000
1964	35,000
1965	37,000
1966	35,000
1967	39,000
1968	32,000
1969	24,000
1970	22,000 [10]

In November 1951 Dinorwic Quarry was incorporated as a limited company with the title the Dinorwic Slate Quarries Company Ltd. The share capital of the company was £850,000, divided into shares of £1 each.[13] In 1962 the company resolved to open a new quarry, as the layout of the existing one was unsuitable for the mechanised methods of production which they wanted to introduce. It was felt also that the new quarry would be able to provide a 'continuation of employment and productive capacity, to offset the running down of Dinorwic Quarry'.

The new Marchlyn Quarry was situated at a height of 1,500ft on the western slopes of the Elidir Mountain. Excavations had been made as early as 1931; but they were interrupted by the outbreak of war and were not renewed on a large scale until 1961, when large quantities of boulder clay were removed by traxcavators and Foden dumpers.[14] Despite the large sums invested, the Marchlyn Quarry proved a failure.

In 1962 the Dinorwic Quarry tried to develop slate as a flooring material; and this proved more successful, as the floors were hard-wearing and resistant to rising damp. Other products included mural slates, slated coffee-tables, and fireplace units. Dinorwic coffee-tables were available only in limited quantities. Specially selected flawless slabs were used, and these were hand processed and wax-finished, the colours being 'sage green or rose grey'. Fireplaces in 'sea green mottled slate' were also available.[15]

In 1962 the annual output of Dinorwic Quarry was about 1,000 tons; and about 700 men were employed. In the same year Dorothea Quarry, with the 'biggest pit in Britain', set out to become a tourist attraction. After paying an admission charge, visitors were taken round the quarry and the dressing sheds; and the highlight of the tour was the horse which pulled the slate wagons in and out of the sheds.[16]

On 1 July 1967, Dinorwic Quarry reduced by 5 per cent and 15 per cent the prices for slates measuring 24in × 12in and 20in × 10in. The new prices per mille of 1,000 slates were as follows:

	Best	*Seconds*	*Thirds*
24in × 12in	£169 5s	£163 19s	£147 16s
20in × 10in	£99 3s	£96 4s	£86 11s

This high price meant that the private housing market preferred concrete tiles to slates. In November 1965, however, Dinorwic Quarry was producing an average of 9,000 tiles and 15,000 bricks per day; both bricks and tiles were made from the small waste which was processed into crushed aggregates.

The possibilities of the export market were explored by the Dinorwic Company in 1965; and in June a representative of the company visited the old slate quarries of Ardoisières d'Angers, where about 3,000 men were employed on the mining of slate. A market for Dinorwic slate was found there; and an agreement was made to supply the French firm with quarterly cargoes of about 300 tons each. These were to be mainly of red slates measuring between 16in × 12in and 10in × 6in; and any available purple slates were to be included.

The freight charges to St Malo were about £3 5s per ton; and this compared 'very favourably' with the charges for slates sent by road to London.[17] Dorothea Quarry exported flooring slates to France; and in 1968 the demand was said to be growing rapidly.[18]

In 1972 Penrhyn Quarries Ltd were producing 11,000 tons of roofing slates annually. The modernisation programme, which began in 1964, included the establishment of new splitting, sawing and architectural departments; the last of the old splitting sheds was closed in 1967. The architectural department was producing 2,000 tons in 1972; and in 1966 frame sawing machines were introduced for sawing blocks in this department. In 1966 Wessex diamond saws were introduced for sawing slate blocks for the production of roofing slates. The Fullersite plant has been modernised, and is now producing 300–400 tons per week of fine inert slate powders for the plastic industry and other industries. A new granule plant now produces slate materials used in the production of roofing felts and other products. The markets for slates from Penrhyn include the United Kingdom, Ireland, France, Holland, Denmark, Belgium, Australia, the USA, New Zealand and Austria. The slates are sold per thousand, and are sent by road for the home market and by sea, from either Port Penrhyn or Port Dinorwic, for export orders.

In 1969 Dinorwic Quarry employed about 300 men; but in July of that year the quarry closed down. The immediate reason for the closure was the delay in receiving French orders. French customers were using up their own stocks; and this resulted in a serious shortage of cash for the firm. In addition, the amount of private building at home was very small; and Dinorwic Quarry was old, and no longer capable of producing slates on a commercial basis. In 1972 the quarry's workshops were converted into the North Wales Quarrying Museum; and along the shores of Llyn Padarn a narrow-gauge railway was laid down for the benefit of tourists. The quarries themselves are owned by Penrhyn Quarries Ltd, but are not worked.[19]

In 1970 the Dorothea Quarry Company announced its closure. The reasons given included the decline in house building since

1969; and it was said that the government's credit squeeze had put too great a strain on a small company in process of modernising its methods of production.[20] In 1972 the quarry property, which extended over 1,000 acres, was sold to Dafydd Gwyn Evans; it is now being developed as a business venture connected with the tourist industry. The Votty and Bowydd Slate Quarries closed in January 1963; and the Oakeley Quarry closed in June 1971. The Oakeley Quarry was the largest slate mine in the world; and it contained some 50 miles of railways, mostly underground. The depth of working was over 1,400ft; and this entailed the use of powerful hydraulic and electric pumps.

In Blaenau Ffestiniog local residents and the Welsh Tourist Board subscribed to a business venture called 'Quarry Tours', which opened its first ½ mile of tramway in the Llechwedd slate mines early in 1972. The tramway has battery-operated electric locomotives, which haul passenger cars along part of the original level opened in 1846. Tools and equipment are on display to show how the slate was won, and the tourists can also watch a splitter at his traditional task. Plans exist for linking Quarry Tours' track with the Festiniog line, which now finishes about 4 miles from Blaenau Ffestiniog.[21]

The failure of local authorities to use slates in their housing programmes has been much criticised by quarry proprietors. Local authorities in Wales are eligible for a subsidy if they use slates, but they rarely take advantage of this arrangement. The majority of new houses in Caernarvonshire and Merionethshire, however, are roofed with slate. The market in slates for roof repairs is contracting, because when clearances are made, the tendency is to replace them with flat-roofed multi-storey buildings. Since the 1960s there has been a rapid growth in the architectural use of Lake District slate; but lack of capital has prevented Welsh firms from competing in this market.

Penrhyn Quarry continues to dominate the North Wales slate industry. The other firms which were still producing roofing slates in 1972 included Maenofferen Slate Quarry Company Ltd. At that time the company employed sixty men; and was producing an annual output of about 1,100 to 1,200 tons, a quarter of which was exported. Of the quarries organised by the Caernarvonshire Crown Slate Quarries Company Ltd, only Moeltryfan Quarry still functioned in 1972; and it employed only twelve men.

In 1972, Cwt y Bugail Slate Quarries Company Ltd employed four men in their quarry; and Penyrorsedd Slate Quarry Company Ltd employed about twenty. The Aberllefenni Slate and Slab Quarries Company Ltd, now known as Wincilate Ltd, employed thirty men in the quarries of Aberllefenni and Braich Goch, and about twenty men at Caernarvon where the slabs for sills and cladding were finished. Their average annual output is about 1,260 tons.[22]

Some writers have complained about the 'lamentable eyesore that such great quarrying centres as Ffestiniog and Bethesda form upon the fair face of Wales.'[23] Others have considered Penrhyn Quarry a 'prodigious chasm' affording 'eminent proofs and effects of those great and useful powers of mind and body, which the Deity has bestowed on man'.[24] It cannot be denied that the products of the North Wales slate industry provided a cheap, durable, weathertight roofing material for millions of homes. It may also be claimed that they added interest to the British landscape. Against the rage of William Morris can be set such tributes as the following:

> They reflect, like the waters of a lake, the changing hues of the sky, causing a roof of Portmadoc Slate to vary in colour from a deep inky blue to a hazy grey.[25]

GAZETTEER

This is a list of the more important quarries. A complete list would be much longer.

ABERCORRIS

SH 754085. Situated in the Corris area, and owned by the Abercorris Slate Company Ltd. In 1903 the annual output was 1,167 tons; later it amalgamated with Braichgoch Quarry.

ABERCWMEIDDAU

SH 746089. Located in the Corris district. The quarry was open, and was served by a branch of the Corris Railway, which ran from Maespoeth via Braichgoch to the Abercwmeiddau workings, and was worked by horses. The slate and slabs from the Broad Vein were taken down to the river at Machynlleth and put in small boats. Sometimes the cargoes were transhipped into seagoing ships at Aberdovey. The slate was usually very hard, and did not cleave into thin sections. This made it more expensive to produce. In 1883 the annual output was 2,875 tons, and the number of men employed was eighty. The quarry closed about 1890.

ABERGYNOLWYN OR BRYNEGLWYS AND CANTRYBEDD

SH 695054. Situated in the Corris district at Bryneglwys. In 1847 John Pugh began quarrying on the Hendre estate, and the slate from the Narrow Vein was carried over the mountains by packhorse to Aberdovey, where it was loaded into ships. In 1864, William and Thomas Houldsworth McConnel, proprietors of cotton mills in Manchester, formed the Aberdovey Slate Company, and leased land at Cantrybedd, Bryneglwys and adjoining properties to work the slate. They promoted the Talyllyn Railway, which was opened in 1865. In 1867 the concern became known as the Abergynolwyn Slate Company Ltd. The slate was worked underground. A vertical shaft almost 200ft high contained two hoists for the slate, and this was powered by two 30ft waterwheels, fed from a reservoir higher up the hill. From the foot of the shaft a drainage tunnel ran north till it reached the surface at the top of the Cantrybedd level. Three tunnels connected with the shaft; all were large enough to walk through, and tramways were laid through them. By 1879 260 men were employed; and there were sixteen chambers, each about 120ft × 60ft. An inclined plane replaced the hoists in the original shaft. In 1883 Abergynolwyn Company went into voluntary liquidation; and the McConnel brothers purchased the leases, part of Abergynolwyn village and the shares in the Talyllyn Railway Company. In 1895 the property was organised by William Houldsworth McConnel junior; and in 1911 it was sold to Henry Haydn Jones, MP for Merioneth. He reopened the quarry, which had been closed in 1910. In 1937–8 there were fifty-seven employees, and the annual output was 88 tons. The use of slate from supporting pillars led to the collapse of the best workings in the centre of the Narrow Vein in 1946; and in 1948 the quarry was closed. The Talyllyn Railway continues to run as a tourist attraction.

ABERLLEFENNI

SH 768103. Situated in the Corris district, about 2 miles up the Dulas Valley. The quarry dates from about 1500, but it was revived around 1810. The slate was worked underground from the Narrow Vein, and it was transported by a tramway connected with the Corris Railway. This line was opened as the Corris, Machynlleth & R. Dovey Tramway in 1859, and was worked by horses until 1879. Aberllefenni was really three quarries: Ceunant Du, Hen Chwarel and Foel Grochan. Slabs only were obtained from Foel Grochan, as the rock was too hard to produce roofing slates. In 1878 180 men were employed; in 1937–8 there were 134 employees; in 1946 there were seventy; and in 1972 Aberllefenni and Braichgoch Quarries were worked under one management, Wincilate Ltd, and employed a total of about thirty men. They produced only slabs.

ALEXANDRA

SH 519569. Situated on Moeltryfan Mountain. A company was formed in 1862, with a capital of £15,000, and 140 men were employed. The annual output in 1874 was 6,000 tons, and a new company was formed. In 1918 the quarry was taken over by the Caernarvonshire Crown Slate Quarries Company Ltd. The quarry was no longer worked in 1972.

BLAENYCAE

SH 485538. The quarry was in Nantlle; and in 1905 it was organised by Thomas Robinson, who owned Talysarn, Tan'rallt, Cloddfa'r Coed and Braichmelyn. The quarry was then acquired by the Dorothea Slate Company, but it was not worked by them. It was closed about 1909.

BLAEN Y CWM

SH 739460. This was an open quarry in Penmachno in the Ffestiniog slate range. It was opened about 1820, and was auctioned in 1826, along with Tanyrhiw Slate Quarry. Both these quarries were situated on farms, from which they got their names. Blaen y Cwm was being worked in the 1860s, but it closed about 1900.

BOWYDD

SH 725468. The quarry, an underground working in the Ffestiniog district, dated from about 1801, when it was worked by Lord Newborough. In 1846 it was revived by F. S. Percival; and in 1875 it amalgamated with the Votty Quarry. The two quarries were separated by a geological fault, and Votty was utilised as a large reservoir. The concern became a limited liability company; and in 1900 the mine and mills were electrified by the Yale Electric Power Company Ltd. In 1903 the annual output was 11,345 tons, and about 366 men were employed. F. S. Percival died in 1913; and the company amalgamated with the Oakeley Company in 1933. The mine was closed in January 1963.

BRAICH OR OLD BRAICH

SH 509559. Situated on the brow of the Cilgwyn Mountain in Nantlle. The quarry dated from 1830, but was reopened in 1860 by a private company, and in 1873 140 men were employed. The quarry consisted of one large pit with three floors. The slates were transported by means of a tramway laid along the side of the Cilgwyn Quarry and connected with the London & North Western Railway. In 1882 the company amalgamated with the Fron or Vron Quarry to form the Fron and Old Braich Welsh Slate Quarries Company Ltd, and the authorised capital was £15,000. The quarry closed about 1914.

BRAICHGOCH OR BRAICH GOCH

SH 748078. The quarry was in the Corris district, on the southerly side of the pass from Talyllyn, and dated from about 1830. It was revived about 1870 by a private company, and 200 men were employed. The Narrow Vein was worked to produce slabs and roofing slates, which were transported by the Corris Railway to Machynlleth. In 1903 Braich Goch and Gaewern Quarries together produced 5,086 tons, under the title the Braichgoch Slate Quarry Company Ltd. This had fallen to 1,275 tons in 1907. In 1972 the quarry was one of the Wincilate Group; and modern plant and equipment had been installed for the extraction of slate blocks, including a specially constructed chain-cutting machine which cut the rock vertically and horizontally in order to extract blocks of pre-determined size of up to 4 tons in weight.

BRYNEGLWYS AND CANTRYBEDD

See *Abergynolwyn*.

BRYN HAFOD Y WERN

SH 631693. Situated about 3½ miles south-east of Bangor, at the foot of Moel Wnion in the parish of Llanllechid. The quarry was worked in the eighteenth century and was leased from the Crown by the Pennants. In 1817 George Hay Dawkins Pennant obtained a new lease, but in 1825 he gave up working the quarry. In 1845 the Royal Bangor Slate Company worked the quarry, and slates were sent by cart to Bangor. The annual output in 1856 was 4,379 tons, but this fell to 2,198 tons in 1882, and the quarry failed to prove an economic proposition. It was dependent on the goodwill of the Pennants for its water-supply, and in 1884 the supply was cut off. The quarry was then abandoned, and now only rubbish tips and a large waterhole remain.

BRYN MAWR

SH 555599. The quarry was at Llanberis. In 1883 the annual output was 160 tons; and in 1886 the quarry was purchased by the Llanberis Slate Company Ltd, which went into liquidation in 1928.

BUGAIL AND CWT Y BUGAIL

SH 733471 and 734468. The Cwt y Bugail Slate Company Ltd, in Penmachno, employed over 100 men in 1873, some of whom lived in well-run barracks. The company provided a small library for the men, and allowed them to have a night school in the manager's office. In 1882 the annual output was 2,549 tons and seventy-three men were employed. By 1937–8 there were forty-two men working; and in 1972 only four men were employed. The concern was two undertakings: a large open pit and an underground working.

BWLCH SLATERS OR BWLCH Y SLATERS

See *Manod.*

BWLCH Y DDEILOR OR BWLCH Y DDWY-ELOR

SH 555509. Situated at Rhyd-Ddu. The annual output was 160 tons in 1883 and the number of men employed was seven. The quarry was closed about 1890.

BWLCH Y GROES

SH 560600. Situated at Llanberis; in 1886 the quarry was purchased by the Llanberis Slate Company Ltd, which was wound up in 1928.

CAERMEINCIAU

SH 566603. Situated 1 mile west of Llanberis; and owned in 1882 by J. T. Campbell. The output in that year was 240 tons, and thirteen men were employed. The quarry was closed about 1900.

CAMBRIAN

See *Glyn Ceiriog*.

CAMBRIAN

SH 560600. Located to the south-west of Llanberis, and owned by the Cambrian Slate Quarries Company Ltd in 1878, when 120 men were employed. The output at that date was 5,000 tons. The slates were red and purple-spotted. The quarry was worked, with the Goodman Quarry, by T. H. Goodwin Newton; and in 1878 these two quarries amalgamated with Cefn Du Quarry. The Cambrian, Goodman and Cefn Du Quarries were auctioned in 1930.

CAMBRIA WYNNE

SH 723058. The quarry was between Corris and Machynlleth, and dated from about 1873. It was worked in galleries, and fifty men were employed. The output was almost wholly slabs, which were sent by tramway to Machynlleth. The quarry closed about 1890.

CAPEL CURIG

See *Rhos*.

CARLISLE

See *Dinas Mawddwy*.

CEDRYN

SH 729632. Situated in the Conway Valley. Lord Newborough leased the quarry to Richard Griffith and W. A. Darbishire in the 1860s. The lessees constructed a railway from the quarry to Borthllwyd on the River Conway. The quarry was closed about 1890.

CEFN DU

SH 555604. The quarry was in the parishes of Llanrug and Llanbeblig. It was leased from the Crown in 1800 by John Evans and four partners for thirty-one years, and was subsequently sub-let. In 1827 the quarry was up for auction, after the death of John Evans. From 1875 the quarry was worked by James Wotherspoon; and in 1878 it was one of three worked by the Llanberis Slate Company Ltd. The output in 1882 was 5,640 tons, and 197 men were employed. In 1883 Chwarel Fawr, adjoining Cefn Du, was purchased; and in 1886 Bryn Mawr and Bwlch y Groes Quarries were purchased. The company went into liquidation in 1928; and in 1930 the quarry plant was dismantled and auctioned. The Cefn Du auction list included store houses, a powder magazine, a sheltering shed, a miners' hut and a dinner shed.

CILGWYN

SH 498538. The quarry was on the Cilgwyn Mountain, in the parishes of Llandwrog and Llanwnda. It dated from the twelfth century and was reputed to be the oldest quarry in Wales. In 1745 John Wynn of Glynllifon leased the quarry from the Crown for thirty-one years. In 1827 the quarry was auctioned; and in 1878 Hayward & Company worked it. The output in 1882 was 7,430 tons, and 300 men were employed. There were four pits: Faengoch, Old Cilgwyn, Cloddfa'r Dwr and Cloddfa Clytiau. The quarry was amalgamated with four others in the Caernarvonshire Crown Slate Quarries Company Ltd in 1918; but by 1972 it had ceased to be worked.

CLODDFA'R COED OR GLODDFA GOED

SH 493529. Situated in the Nantlle Valley. The quarry was opened about 1790; and reopened in 1870. In 1873 only four men were employed. Ten men were employed in 1937–8; and in the 1940s the quarry was acquired by the Dorothea Slate Quarry Company Ltd, but not worked.

CLOGAU OR CLOGIE

SH 185463. Located in Denbigh. In 1937–8 nineteen persons were employed by the Berwyn Slate Quarries Company Ltd. The quarry was closed by 1946.

COED MADOG

SH 491529. Situated in the north-east of the Nantlle Valley. In 1882 the annual output was 1,871 tons and 100 men were employed by the Coed Madog Slate Company. The quarry was an open pit; and was abandoned in 1909.

COETMOR

See *Pantdreiniog.*

COOK AND DDOL

SH 560605. The quarries were south-west of Llanberis in the parish of Llanrug. They were leased by John Evans and his partners from the Crown; and they were sub-let by them in 1812 to four partners for fifteen years. One of the partners was Thomas Bulgin, a Bangor slate merchant, and the others included a Chester stationer and a local quarryman. In 1813 the output was 500 tons; and in 1820 Thomas Bulgin and two partners worked the quarries for a rent of 5s a year and one-ninth of all slates made. This partnership was dissolved in 1823. The quarries were worked by Trew Iegon in 1873, and forty men were employed. In 1882 the number of men was reduced to twenty-six, and the annual output was 616 tons. In 1905 the quarries were purchased by the Co-operative Quarries Ltd; but they were up for sale in 1911. The quarries closed by 1937.

CORWEN OR PENARTH

SJ 107424. Situated near Corwen. In 1883 ten men were employed and the annual output was 498 tons. In 1903 the quarry was worked by the Corwen Slate Mining Company; and the annual output was 1,707 tons. The quarry was served by the Penarth Tramway, and was closed about 1920.

CRAIG DDU OR GRAIG DDU

SH 724454. Situated in the Ffestiniog district. The quarry was an open working, dating from 1840. In 1873 106 men were employed; and it was worked by a limited liability company. In 1883 the annual output was 807 tons and thirty-six men were employed. A tramway was laid from the summit to the road in a series of four inclined planes, and the men used to travel down these in trolleys or 'wild cars'. In 1937–8 eighty-six men were employed in the quarry. It was closed by 1946.

CROESOR

SH 657457. The quarry was on Moelwyn Mawr. It dated from 1846; and from 1866 to 1876 it was worked by the Croesor

United Slate Company. It was an underground working. In 1862 the output per annum was only 74 tons; but this was increased to 1,190 tons in 1864. In 1865 the Croesor & Portmadoc Railway Act incorporated the Croesor & Portmadoc Railway Company, to maintain an existing railway from Carreg Hylldren to Portmadoc. In 1875 the Croesor New Slate Company Ltd acquired the concern, but by 1882 they had given up the quarry. In 1895 the Park and Croesor Quarries were amalgamated to form the Park & Croesor Slate Quarries Company Ltd, with the head office at Penrhyndeudraeth near Portmadoc. The authorised capital was £100,000. The annual output was low, being about 24 tons; and in 1914 a receiver was appointed. The quarries finally closed in 1930.

CWM EIGIA

SH 708635. Situated in the Conway Valley. The quarry was leased from Sir R. B. W. Bulkeley about 1860 by a company which had an authorised capital of £30,000. In 1867 the annual output was 588 tons of roofing slates and 140 tons of slabs. The quarry was closed about 1890.

CWM ERA

SH 760064. This was an underground concern, situated in the Corris district, which dated from about 1870. It was abandoned about 1900.

CWM MACHNO OR PENMACHNO

SH 754472. Situated in Penmachno. In 1882 the annual output was 4,412 tons, and 100 men were employed. In 1946 the annual output was 1,595 tons and seventy-five men were employed. It was closed by 1972.

CWMORTHIN OR CWM ORTHIN

SH 689469. The quarry was opened about 1820, and was in the valley on the eastern margins of the Moelwyns. Thomas Casson and two partners first began quarrying, but in 1830 the quarry was abandoned. It was reopened about 1845. In 1889 the New Welsh Slate Company Ltd was formed, largely financed by English capital; and the nominal capital was £65,000. The number of men employed in 1893 was 291; and in 1895 the annual output was 7,252 tons. In 1896 the quarry was operating on co-operative lines: one-tenth of the profits was divided as a dividend among the workmen who held £5 in shares in the company, and a further one-tenth of the profits was divided among those who held £100 in shares. In 1900 the company was acquired by the Oakeley Company.

CWM TRWSCWL

See *Prince of Wales.*

CWT Y BUGAIL

See *Bugail.*

CYMERAU

SH 777107. Situated north of Corris. In 1883 the output was 762 tons, and twenty-nine men were employed. In 1937 the quarry was worked by Jones, Indigo & Company, and six men were employed. The quarry was closed by 1972.

CYNICHT

SH 646467. Situated on the Cynicht Mountain in the Ffestiniog district. It dated from about 1876, although in 1845, when the ground was auctioned, the slate rocks and mineral veins were said to be partly worked. The quarry closed about 1900.

DINAS MAWDDWY OR MINLLYN OR CARLISLE

SH 858141. Situated on the slopes of a hill called Bryn-yr-Wylfa in the Dovey Valley. The quarry was worked mainly for slabs. It was connected by the Dinas Mawddwy Branch Railway with the Cambrian Railways at Cemmes Road. The line was of standard gauge, and was owned by a private company. Sir Edmund Buckley was the principal shareholder. In 1883 the annual output was 1,468 tons. The quarry was closed in 1915.

DINORWIC

SH 595605. Situated at Llanberis on the Elidir Mountain. In 1787 the old quarries which had been opened by local quarrymen were leased by Thomas Wright, Hugh Ellis and William Bridge, from Thomas Assheton Smith, the lord of the manor. In 1809

Thomas Assheton Smith along with Thomas Wright, William Turner and Hugh Jones worked the quarry. In 1824 a railway was built to Port Dinorwic; and in 1841–2 a new railway was made. In 1870 new workshops were erected in a large rectangular building, and these were converted into a quarry museum in 1972. In 1882 the annual output was 87,429 tons, and the number of men employed was 2,757. The quarry was worked in galleries, and there were sets of inclined planes ascending to the summits of the quarries called Garrett, Wellington and Victoria. In 1968 about 300 men were employed. The quarry closed in 1969; it is now owned, but not worked, by Penrhyn Quarries Ltd.

DIPHWYS CASSON

SH 712467. Situated at Duffws in the Ffestiniog district. It was opened about 1765; and in 1800 William Turner and William and Thomas Casson worked it. They constructed a road from the quarry to Congl-y-Wal in 1801. The number of men employed in 1901 was 196; and during World War I the quarry closed. It was reopened in 1920, but closed again in 1927. In 1936 it re-opened and became a subsidiary of the Oakeley Company. The quarry closed in 1955.

DOROTHEA

SH 499532. Situated in the centre of the Nantlle Valley. The original proprietor was Richard Garnons, and the quarry was first opened by William Turner and John Morgan about 1829. The quarry was served by the Nantlle Railway and later by the London & North Western Railway. For a few years, from 1848, the quarry was worked by a group of Nantlle workmen. In 1853 these workmen sold their shares to John Williams, who eventually became the chief shareholder. The Williams family maintained a controlling interest in the quarry until its closure in 1970. There were four open pits; and in 1882 the annual output was 16,598 tons, with 533 men employed in the quarry. The company acquired Penybryn Quarry in 1894; and in 1899 the River Llyfni was lowered and diverted so that it was nearer Nantlle Lake, which was drained. In 1921 the company purchased South Dorothea Quarry; and in 1933 Galltyfedw was worked by the company. The site, in 1972, was being developed as a tourist attraction.

DRUM

SH 735431. Situated to the north of Llyn-y-Drum in the Ffestiniog district. The quarry was an open working, which dated from about 1864. It was closed about 1880.

EUREKA

See *Llwydycoed.*

FOEL FORFYDD OR FOEL FAEN

SJ 187477. Situated in Denbighshire. In 1937 the quarry was run by J. E. Baines; six men were employed. It was closed by 1946.

FRON OR VRON

SH 514549. Situated in the north-eastern extremity of Nantlle. It dated from about 1830, and was revived in 1860. In 1872 the annual output was 1,500 tons; and in 1873 the number of men employed was eighty. There was an incline down from the Cilgwyn Tramway, to carry slates from Fron to the Nantlle Railway, but this was not used after 1881. In 1882 the quarry amalgamated with the Old Braich Quarry, and at that date sixty-two men were employed. In 1937 the quarry was worked by O. J. Hughes & Son, and seven men were employed. The quarry closed about 1950.

FRONHEULOG OR VRONHEULOG OR NEW VRONHEULOG

SH 489519. Located in the southern part of the Nantlle Valley. The quarry dated from about 1840, and was revived in 1866 by William Turner & Company Ltd. In 1873 twelve men were employed. The slates were sent to the Nantlle Railway along a tramway running from Tan'rallt Quarry. In 1882 the quarry was worked by the New Vronheulog Slate Quarry Company Ltd; and the annual output was 1,642 tons. The number of men then employed was ninety-eight. The quarry was closed by 1914.

GAEWERN

See *Garwen.*

GALLTYFEDW

SH 499535. The quarry was in the Nantlle Valley, between Dorothea and Cilgwyn Quarries. There were two pits, one of which was flooded in 1873. The quarry was reputed to be one of the oldest in the valley. In 1933 the quarry was acquired by the Dorothea Slate Company, but was not worked.

GALLT Y LLAN

SH 627543. Situated at Llanberis. The quarry dated from 1811, when it was leased by John Evans's partner to a quarryman and a slate loader. In 1882 the annual output was 90 tons and three men were employed. In 1937–8 three men were employed; and the quarry was closed by 1946.

GARREG FAWR

SH 538584. Situated at Waunfawr. The quarry dated from 1802, when it was leased by John Evans and his partners to a group of yeomen and quarrymen. In 1882 it was worked by the Bettws Garmon Slate Company. At that date the annual output was 80 tons and six men were employed. It closed about 1900.

GARWEN OR GAEWERN

SH 745086. Situated in the north-east of the Corris district. The quarry was an underground working, and dated from about 1820. It was revived in 1858; and in 1873 the Talyllyn Slate Company worked the quarry and over 200 men were said to be employed. In 1884 the quarry was acquired by the Braichgoch Slate Quarry Company Ltd; and in 1972 this company formed part of the Wincilate Group.

GLANRAFON

SH 581540. Located near Rhyd-Ddu. The quarry was worked by the Glanrafon Slate Quarry Company Ltd. In 1882 the annual output was 992 tons, and ninety-seven men were employed. The company was registered in 1904, with an authorised capital of £50,000. The quarry was closed about 1914.

GLODDFA COED

See *Cloddfa'r coed.*

GLYN CEIRIOG OR CAMBRIAN

SJ 189388. Situated in Denbighshire. The quarry dated from the seventeenth century. It was served by the Glyn Valley Tramway, of 2ft 4½in gauge. In 1903 the annual output was 2,210 tons. It was closed by 1937.

GLYN CEIRIOG OR WYNNE

SJ 187377. Situated in Denbighshire. The quarry was also served by the Glyn Valley Tramway; and in 1903 the annual output was 2,166 tons. The quarry was closed by 1937.

GRAIG DDU

See *Craig Ddu.*

GLYNRHONWY LOWER

SH 570610. Situated on the Glynrhonwy estate belonging to Lord Newborough, on the western side of Llyn Padarn. The quarry dated from the early eighteenth century; and in 1858 it was referred to as 'Captain Taylor's Quarry' in an advertisement inviting offers for a partnership. In 1872 the annual output was 8,000 tons, and 200 men were employed. In 1882 the annual output was 1,514 tons, and the number employed was fifty-three. The quarry closed about 1930.

GLYNRHONWY UPPER OR PREMIER GLYNRHONWY OR UPPER GLYNRHONWY

SH 566609. Situated at Llanberis. In 1873 the number of men employed was forty. In 1882 the annual output was 2,181 tons, and ninety men were employed. In 1919 the Premier Glynrhonwy Slate Quarries Company Ltd was formed to acquire the Upper Glynrhonwy Slate Company Ltd.

GOODMAN

SH 566605. Situated at Llanberis. In 1873 the concern was worked by a private company, and twenty men were employed. In 1878 the company amalgamated with the Cambrian and Cefn Du Quarries to form the Llanberis Slate Company Ltd. The lessor was Bishop Goodman's trustee; and the area covered about thirty-five acres. The quarry ceased to be worked about 1890; and it was dismantled in 1930.

GORSEDDAU

SH 574454. Situated in the parish of Penmorfa, in the Pennant Valley. The company working it began operations in 1855. In 1859 the annual output was 1,338 tons, and 200 men were employed. A railway was constructed from the quarry; and in 1872 this was incorporated in the Gorseddau Junction & Portmadoc Railway Company. The quarry went into liquidation in 1870; and in 1872 it was purchased by the Prince of Wales Company. By 1882, however, both quarries were closed. A ruined slate mill about 2 miles from the quarry can still be seen.

GREAVES

See *Llechwedd.*

GWERNOR

SH 499528. The quarry was situated about 6 miles south of Caernarvon. In 1873 there were twenty men employed and the annual output was 1,000 tons. The quarry produced 'silver green' coloured slates, for which there was a great demand. It was not, however, possible to work the green rock without working the blue, and from about 1910 this was a depressed trade. Water power was used for haulage, pumping and slate dressing purposes. The quarry closed about 1930; it was later acquired, but not worked, by the Dorothea Slate Quarry Company Ltd.

HAFODLAS

SH 489539. Situated in the Nantlle Valley. The quarry dated from about 1800; it was one of the earliest to employ a steam engine for pumping purposes, although the engine was broken in 1817. George Bettiss was one of the partners at that date. The quarry closed about 1880.

HAFOD Y LLAN OR SOUTH SNOWDON

SH 617513. Situated in the Gwynant Valley. The quarry dated from 1840, and was worked by a London company. It closed about 1880.

HAFOD Y WERN

SH 532572. Situated near Bettws Garmon. In 1882 it was

owned by the Moel Tryfan Crown Slate Company Ltd. It closed about 1885.

HENDDOL

SH 621121. Situated at Ffriog, south of Dolgellau. In 1883 the annual output was 401 tons, and the number of men employed was forty. The quarry had a tramway, $\frac{1}{4}$ mile long, with a 1ft 11$\frac{1}{2}$in gauge. The quarry closed about 1920.

HENDRE

SH 745526. Located near Dolwyddelan. The quarry was worked by the Hendre Slate Quarry Company Ltd. The annual output was 120 tons in 1882, and six men were employed. The quarry closed about 1890.

HENDRE DDU

SH 798126. Situated in the Corris district. The quarry was open, and was owned by Sir Edmund Buckley. In 1905 the annual output was 700 tons, and it was worked by Jacob Bradwell. The quarry, and those around it, were served by the Hendre Ddu Tramway, which had a 1ft 11$\frac{1}{2}$in gauge, and was worked by horses. The line connected with the Mawddwy Railway at Aberangell, but by 1939 it had been taken up. In 1937–8 the quarry was worked by T. Glyn Williams & Company, and seventeen men were employed. The quarry was closed by 1946.

HOLLAND'S OR RHIWBRYFDIR OR UPPER QUARRY

SH 691470. Located in the Ffestiniog district. The quarry was worked by Samuel Holland from about 1825. Holland's Quarry was approached by a series of inclines; and the workings were both open and underground. In 1873 500 men were employed; and in 1878 the quarry was taken over by W. E. Oakeley. It was worked by him until 1882, when a company was incorporated to work the quarry.

ISALLT

SH 533445. Situated in the Pennant Valley. The quarry was opened about 1840, and was closed about 1860.

LLANRHYCHWYN

SH 763608. Situated in the Conway Valley. The quarry was opened about 1850, and closed about 1880.

LLECHWEDD OR GREAVES

SH 705469. Situated at Blaenau Ffestiniog. The quarry dated from about 1846, when it was worked by John Whitehead Greaves (1807–80). About £30,000 was reputed to have been expended before the valuable Old Vein was discovered. In 1900 the annual output was 23,734 tons, and in 1901 509 men were employed. The quarry was served by the LNWR, the GWR and the Cambrian Railways. In 1972 the underground workings consisted of over 25 miles of tunnels connecting vast caverns; and the workings extend to a depth of 900ft. In 1946 the annual output was 3,678 tons and 175 men were employed. In 1972 about seventy men were employed. The original level, opened in 1846, can be visited and toured by underground tram, for about ½ mile. The tramway has battery-operated electric locomotives, which haul specially designed passenger cars.

LLWYDYCOED OR EUREKA

SH 514553. Situated in the Nantlle area, in the parish of Llanllyfni. The quarry dated from about 1850. In 1882 it was worked by William Jones; the annual output was 32 tons, and six men were employed. The quarry closed about 1890.

LLWYNGWERN

SH 758045. Located in Montgomeryshire. In 1883 the annual output was 915 tons and twenty-six men were employed. In 1937–8 the quarry was worked by the Llwyngwern Slate Quarries Company Ltd, and seventeen men were employed. The quarry closed about 1950.

MAENOFFERN

SH 705466. Situated in the Ffestiniog district. The quarry was opened in 1861 and was worked by a limited liability company. It was an underground working. In 1882 the annual output was 8,600 tons, and 244 men were employed. In 1905 the annual output was 10,984 tons. The company acquired Rhiwbach Quarry in 1928. In 1946 the annual output was 2,852 tons, and 194 men were employed. In 1972 about sixty men were employed and the annual output averaged 1,200 tons.

MANOD OR BWLCH SLATERS OR BWLCH Y SLATERS

SH 735455. Situated in the Ffestiniog district. The quarry dated from 1805, when it was leased to O. A. Poole, one of John Evans' partners. In 1813 the quarry was leased to an architect, Francis Webster, and five others. This early quarry was known as Manod Quarry; and Bwlch Slaters, which was part of this quarry, was leased by John Evans to John Pritchard in 1824. In 1901 fifty men were employed. In 1937 Manod Slate Quarries Ltd worked the quarry, and sixty-six men were employed. In 1954 thirty-five men were employed and the annual output was 470 tons. The quarry was closed about 1960.

MIDDLE QUARRY

See *Rhiwbryfdir*.

MINLLYN

See *Dinas Mawddwy*.

MOEL FABAN

SH 626678. Situated in Llanllechid parish. The quarry was worked by the Royal Bangor Slate Company about 1850. In 1903 it was acquired by the North Wales Quarries Society Ltd, but in 1911 the quarry was closed.

MOELFERNA AND DEESIDE

SJ 125405 and 138405. Situated near Llangollen. In 1905 the annual output of the quarries was 6,150 tons. In 1946 the annual output was 1,558 tons, and fifty-eight men were employed. The quarries closed about 1960.

MOELTRYFAN

SH 518567. Situated on the Cilgwyn Mountain. The quarry dated from 1800, when it was leased from the Crown by John Evans and his partners. In 1827 the quarry was auctioned. In 1882 it was worked by the Moeltryfan Slate and Slab Company

Ltd. The output at that date was 1,880 tons and eighty-one men were employed. In 1918 the quarry was amalgamated with four others to form the Caernarvonshire Crown Slate Quarries Company Ltd. In 1972 the quarry employs twelve men.

MOELWYN

SH 663443. Situated in the Ffestiniog district. The quarry was worked first by Nathan Rothschild about 1825. In 1860 the Great Moelwyn Slate Company Ltd was formed with the authorised capital of £50,000. In 1874 forty-five men were employed. The quarry closed about 1880.

MOEL-Y-GEST

SH 555388. Situated about 1 mile west of Portmadoc. The quarry was opened about 1780; and slates from the quarry were used by W. A. Madocks in the roofing of houses in Tremadoc in 1805. The quarry closed about 1880.

NANTLLE VALE

SH 497526. Situated at Caernarvon. In 1882 the quarry was worked by the Nantlle Vale Slate Quarry Company. The annual output at that date was 150 tons and the number of men employed was twenty. The quarry was closed about 1910.

NANTGWRYD

SJ 198342. Situated south of Glyn Ceiriog. The quarry was opened about 1790, when it belonged to Richard Myddleton of Chirk Castle. The quarry closed about 1880.

NEW VRONHEULOG

See *Fronheulog*.

NYTH Y GIGFRAN

SH 699459. Situated in the Ffestiniog district. In 1860 a company worked the quarry and built inclines. The quarry was abandoned by 1882; but it was reopened briefly in the 1920s.

OAKELEY

SH 691470. Situated in the Ffestiniog district. The quarry dated from 1755; and in 1878 W. E. Oakeley took over the Upper and Middle Rhiwbryfdir Quarries. In 1888 he acquired the Lower

Rhiwbryfdir Quarry from the Welsh Slate Company. Cwmorthin, Glanypwll, Votty and Bowydd and Diphwys Quarries were all eventually taken over by the Oakeley Quarry, which was one of the largest and most prosperous in the Ffestiniog district. There were 50 miles of railway lines in the mines, and the underground workings extended to a depth of over 1,400ft. There were twelve slate mills worked by electricity. The quarry closed in 1971, but has since re-opened on a small scale.

OLD BRAICH

See *Braich.*

PANTDREINIOG OR COETMOR

SH 623671. Situated at Bethesda. The quarry dated from about 1820. In 1856 it was put up for sale, and advertised as having 'every description of Steam Engine, Plant and Machinery connected therewith and a Water Balance'. The property was held on a lease from the Cefnfaes estate, and in 1903 it was acquired by the North Wales Quarries Society Ltd, along with Moel Faban and Tanybwlch. The quarry was closed in 1911. In 1919 the North Wales Development Company Ltd was floated to produce slate dust products at the quarry; but the concern was financially disastrous, and was wound up in 1926.

PARK

SH 633444. Situated in Cwm Croesor, Merionethshire. The quarry was opened in 1872. It was worked chiefly for slabs and slate ridging, and was an underground working. In 1895 the Park and Croesor Slate Quarries Company Ltd was formed. The quarries closed about 1930.

PENARTH

See *Corwen.*

PENLLYN

SH 732525. Situated in the Ffestiniog district. The quarry was opened about 1870. In 1895 the annual output was 720 tons. The quarry closed about 1920.

PENMACHNO

See *Cwm Machno.*

PENRHYN

SH 620650. Situated on the Elidir Mountain. The quarry dates from the fifteenth century at least. There are twenty-one galleries rising to a height of over 1,200ft; and the quarry is about 1¼ miles in length. The total area of the workings, buildings and tips is 560 acres. There used to be 50 miles of tramway in the quarry, with a 6 mile narrow-gauge railway to Port Penrhyn. These have now been taken up, and replaced by a road from the main splitting area to the highest galleries and from the main splitting area to the bottom gallery. In 1852 a branch of the Chester & Holyhead Railway was opened for the carriage of slate. It was of standard gauge and ran from Penrhyn sidings, west of Llandegai Tunnel, to Port Penrhyn, a distance of 1½ miles. From 1954 it was regarded as a siding; and it was closed in 1963. In 1882 the annual output was 111,166 tons, and 2,809 men were employed. In 1914 1,920 men were employed. In 1946 1,065 men were employed; and in 1972 250 men were employed and the annual output of roofing slates was 11,000 tons. There are also a fullersite plant, a granule plant and an architectural department. Since 1964, Sir Alfred McAlpine & Son have had a controlling interest in the quarry.

PENYBRYN

SH 504533. Situated in Nantlle. The quarry was opened about 1770. In 1882 the annual output was 2,725 tons, and 234 men were employed. In 1894 the Dorothea Company acquired the quarry; and it was worked until 1932.

PENYRORSEDD

SH 518538. Situated in Nantlle. The quarry was opened about 1816 when it was worked by William Turner. In 1854 John Lloyd Jones worked the quarry, and in 1863 it was acquired by W. A. Darbishire & Company. The authorised capital was £20,000, but a great deal of capital was at first expended without adequate return. By 1882, however, the annual output was 7,999 tons, and 261 men were employed. In 1892 the number of men had increased to 445. The concern was noted for its enlightened attitude towards its workers. In 1945–6 the annual output was 3,431 tons, and 180 men were employed. In 1972 about twenty men were employed.

PLAS Y NANT

SH 552562. Situated near Bettws Garmon. The annual output in 1883 was 672 tons, and twenty-eight men were employed. The quarry was closed about 1900.

PREMIER GLYNRHONWY

See *Glynrhonwy Upper*.

PRINCE LLEWELLYN

SH 518449. Situated in the Pennant Valley, near Llanfihangel-y-Pennant. The quarry was opened about 1872 by a private company. It was closed about 1880.

PRINCE LLEWELLYN

SH 749529. Situated east of Dolwyddelan. In 1882 the annual output was 1,865 tons, and seventy-four men were employed. The quarry was closed about 1914.

PRINCE OF WALES OR CWM TRWSCWL

SH 545493. Situated in the Pennant Valley. The quarry was worked in 1873 by a limited liability company, which employed 200 men. The annual output at that date was 5,000 tons. By 1882, however, production had been stopped by a large fall of rock.

RATGOED

SH 784119. Situated in the Corris district. In 1907 the annual output was 621 tons, and in 1937 Ratgoed Quarries Ltd employed seventeen men. The concern was closed by 1946.

RHIWARTH AND WEST LLANGYNOG

SJ 066265. Situated in Montgomeryshire. Rhiwarth Quarry dated from about 1750. In 1937–8 the two quarries together employed twenty-six men. The quarries were closed by 1946.

RHIWBACH

SH 743462. Situated in the Machno Valley. The quarry dated from 1812 and had open and underground workings. In 1928 it was acquired by the Maenofferen Slate Quarry Company.

RHIWBRYFDIR OR HOLLAND'S OR UPPER QUARRY

See *Holland's*.

RHIWBRYFDIR OR MATTHEWS'S OR MIDDLE QUARRY

SH 691470. Situated in the Ffestiniog district, between the Welsh Slate Company's Quarry and Holland's Quarry. In 1838 Nathaniel Matthews acquired the quarry; and in 1878, when the lease expired, the quarry was taken over by W. E. Oakeley. In 1882 the quarry formed part of the Oakeley Company.

RHIWBRYFDIR OR WELSH SLATE COMPANY OR LOWER QUARRY

SH 691470. The quarry, which was in the Ffestiniog district, was opened about 1816. In 1819 Samuel Holland worked it; and in 1825 he and his father sold it to the Welsh Slate & Copper Mining Company, of which Lord Palmerstone was chairman, at a price of £28,000. In 1873 the quarry was reputed to be the largest in Ffestiniog. There were vast caverns, with very broad pillars supporting the roof. In 1873 700 men were employed. In 1883 the Welsh Slate Company surrendered its lease to the Oakeley Company.

RHIWFACHNO

SH 754471. Situated near Penmachno. The quarry dated from about 1838; and it was an open working. It was closed about 1890.

RHIWGOCH

SH 759543. Situated near Dolwyddelan. In 1873 the quarry was worked by a limited liability company. It was worked as an open pit. It closed about 1882.

RHOS OR CAPEL CURIG

SH 718555. Situated in Capel Curig. In 1882 the quarry was owned by the Capel Curig Slate Quarry Company Ltd; the annual output was 1,285 tons, and forty-five men were employed. In 1918 it was amalgamated with the Caernarvonshire Crown Slate Quarries Company Ltd. The quarry was closed by 1972.

RHOSYDD

SH 664461. Situated in the Ffestiniog district. The quarry was opened about 1850; and in 1857 it was acquired by a London company. In 1874 another private company took it over. In 1937–8 the quarry was owned, but not worked, by L. M. and D. M. Colman.

SNOWDON QUARRY

SH 581540. Situated on the westerly slopes of Snowdon. The quarry was opened about 1870. In 1873 about fifty men were employed. The quarry closed about 1882.

SOUTH DOROTHEA

SH 494529. Located in Nantlle. The quarry was opened about 1867. In 1882 it was worked by the South Dorothea Slate Company; and the annual output was 1,040 tons, with seventy men employed. In 1921 the Dorothea Slate Company Ltd purchased the quarry.

SOUTH SNOWDON

See *Hafod y Llan*.

TALYSARN

SH 495534. Situated in Nantlle. In 1802 John Evans and his partners took a lease of three quarries on Talysarn Farm for five years, and the concern was called the Talysarn Slate Company. In 1873 a number of pits formed the Talysarn Slate Quarry Company Ltd, and the company also worked Fron Quarry. In 1882 John Robinson worked the quarries, and 400 men were employed. The annual output at that date was 8,210 tons. In 1904 the company was registered, and was authorised to acquire Talysarn, Braich, Cloddfa'r Coed and Tan'rallt Quarries. The village of Talysarn was the terminus of the Talysarn Railway, which ran from Penygroes to Talysarn. The Talysarn Quarry was acquired by the Dorothea Slate Company Ltd in the 1930s, but it was not worked.

TANYBWLCH

SH 628683. Situated in Llanllechid. The quarry was worked from 1873 by the Port Bangor Slate Company Ltd, but was idle by 1882. In 1903 the quarry was acquired by the North Wales Quarries Society Ltd, and was worked on co-operative lines. The quarry closed in 1911.

TANYRALLT OR TAN'RALLT

SH 491523. Situated in Nantlle. The quarry was first opened in 1820. In 1873 it employed about forty men. The quarry was acquired by the Talysarn Slate Company; but in 1937–8 it was worked by W. R. Morris & Company, and eleven men were employed. The quarry was still in operation in 1972.

TREFLAN

SH 539585. Situated at Waunfawr. The quarry was worked by William Leeming in 1882. It was closed about 1885.

TYDDYN AGNES AND TY-N-LLWYN

SH 488520. Situated in Nantlle. The quarry dated from about 1865. In 1873 about twenty men were employed in the two quarries, and the annual output was 1,000 tons. The quarries closed about 1890.

TY MAWR

SH 495529. Situated in Nantlle. The quarry dated from about 1870. In the 1930s it was acquired by the Dorothea Slate Company Ltd, but was unworked.

TY'N Y COED

SH 652152. Situated at Arthog, south of Dolgellau. The quarry had a tramway of $\frac{3}{4}$ mile with a 1ft $11\frac{1}{2}$in gauge. The quarry closed about 1880.

TY'N FFRIDD

SH 626678. Situated on the Coetmor estate, Bethesda. The quarry was worked in 1805 by John Evans and his partners. The quarry was later acquired by the Penrhyn estate, but was unworked.

TYNYWEIRGLODD

SH 493521. Situated in Nantlle. The quarry dated from about 1860. In 1937–8 the quarry was employing thirty-nine men. It was closed by 1972.

UPPER GLYNRHONWY

See *Glynrhonwy Upper.*

VOTTY

SH 706465. Situated in the Ffestiniog district. The quarry was opened at first by local quarrymen on Lord Newborough's estate, about 1834. In 1875 Votty Quarry was acquired by the company working Bowydd Quarry, and the Votty and Bowydd Slate Quarry Company Ltd was formed. In 1933 the concern was taken over by the Oakeley Company; and Votty and Bowydd Quarries were closed in 1963.

VRON

See *Fron.*

WELSH SLATE COMPANY

See *Rhiwbryfdir.*

WRYSGAN

SH 676458. Situated in the Ffestiniog district. The quarry dated from about 1844. In 1874 thirty men were employed. In 1901 the number had risen to 107. In 1905 the quarry was worked by the Wrysgan Slate Quarry Company Ltd, and the annual output was 2,655 tons. The annual output in 1945–6 was 348 tons, and eleven men were employed. The quarry was closed by 1972.

WYNNE

See *Glyn Ceiriog.*

NOTES

Chapter 1: The Slate Industry to 1731 (pages 11–26)

1 R. Warner, *A Second Walk Through Wales in August and September 1798*, Vol II, pp 185–6
2 W. Bingley, *A Tour Round North Wales Performed during the Summer of 1798*, Vol I, p 185
3 *Archaeologia Cambrensis*, Vol 80, Seventh Series, Vol V, 1925, A. H. Dodd, 'The Roads of North Wales 1750–1850', p 130
4 Ranulf Higden, *Polychronicon*, translated by John Trevisa, Rolls Series No 41, Vol I, 1865
5 R. Warner, *A Second Walk through Wales in August and September 1798*, Vol II, pp 187–8
6 J. Evans, *Letters Written during a Tour through North Wales in the Year 1798 and at Other Times*, p 161
7 Thomas Pennant, *Tours in Wales*, Vol II, p 386
8 *Leigh's Guide to Wales and Monmouthshire*, pp 95 and 9
9 J. Hucks, *A Pedestrian Tour through North Wales*, p 133
10 A. H. Dodd, *The Industrial Revolution in North Wales*, pp 152–61
11 Ibid, pp 169–83
12 *O.E.D.* and F. J. North, *The Slates of Wales*, p 3
13 F. J. North, *The Slates of Wales*, pp 89–90; J. L. Williams and Ifor Williams, *Gwaith Guto'r Glyn*, pp 256–8
14 F. J. North, *The Slates of Wales*, pp 24–7
15 Ibid, pp 38–46
16 B. Smith and T. Neville George, *British Regional Geology. North Wales*, p 82
17 D. C. Davies, *Slate and Slate Quarrying*, pp 30–1, and W. M. Richards, *A General Survey of the Slate Industry of Caernarvonshire and Merionethshire* (unpublished MA thesis), p 3
18 *Royal Commission on the Ancient Monuments of Wales and Monmouthshire*, Vol II, p 263 and lvii

19 University College of North Wales (hereafter UCNW), Bangor MS 8277
20 R. E. M. Wheeler, 'Segontium and the Roman Occupation of Wales', pp 102–4, *Y Cymmrodor*, Vol XXXIII, 1923
21 D. C. Davies, *Slate and Slate Quarrying*, p 161
22 E. Hyde Hall, *A Description of Caernarvonshire* (1809–1811), ed Emyr Gwynne Jones, pp 78 and 286
23 H. T. Riley (ed), *Liber Custumarum Compiled in the Early Part of the Fourteenth Century*, Part I (London 1860), pp xxxi–xxxii.
24 *The Transactions of the Society of Cymmrodorion 1915–16* (London 1917), C. R. Peers, 'Caernarvon Castle', pp 17–25
25 *Ministry of Works Official Guide Book*, A. J. Taylor, 'Conway Castle and Town Walls' (HMSO 1961), pp 6–7, 14 and 47
26 H. J. Hewitt, *Medieval Cheshire* (Manchester 1929), pp 89, 100
27 J. Evans, *A Tour through Part of North Wales in the Year 1798 and at Other Times*, p 260
28 Douglas B. Hague, 'The Bishop's Palace, Gogarth, Llandudno', p 14, *Transactions of the Caernarvonshire Historical Society* (hereafter *TCHS*), Vol 17, 1956
29 John Webb (translator), 'An Account of the Treachery of the Earl of Northumberland and the taking of His Majesty Richard II, his Progress from Conway to Rhuddlan, Flint and Chester by an Eye-Witness' (1824). Quoted in W. Bezant Lowe, *The Heart of Northern Wales, As It Was, and As It Is* (Llanfairfechan 1912), Vol I, p 201
30 F. J. North, *The Slates of Wales*, p 89 and Robert Williams, *The History and Antiquities of the Town of Aberconwy*, p 87
31 L. F. Salzman, *Building in England down to 1540* (Oxford 1952), p 233 and E. M. Jope and G. C. Dunning, 'The Use of Blue Slate for Roofing in Medieval England', *The Antiquaries Journal*, Vol 34, July 1954, pp 210–11
32 Quoted in D. C. Davies, *Slate and Slate Quarrying*, p 162
33 George Owen of Henllys, *The Description of Pembrokeshire*, ed Henry Owen (London 1892), p 81
34 W. J. Loftie, *Views in North Wales*, pp 90 and 71
35 F. J. North, *The Slates of Wales*, p 73 and *Penrhyn Slates* (Penrhyn Quarries leaflet)
36 *Calendar of Wynn of Gwydir Papers, 1515–1690 in the National Library of Wales and Elsewhere* (London 1926), nos 61, 185, 2,132
37 *O.E.D.*
38 J. Leland, *The Itinerary in Wales of John Leland in or about the years 1536–9*, ed L. T. Smith, pp 75 and 94
39 G. D. Owen, *Elizabethan Wales*, p 158
40 *Bye-gones*, 1895–6, Vol IV, p 203

41 *Bye-gones*, 1899–1900, Vol VI, p 16
42 *Royal Commission on Ancient Monuments in Wales and Monmouthshire*, Vol II, pp 42, 98 and 172
43 Thomas Pennant, *Tours in Wales*, Vol II, pp 386 and 196
44 David Bremner, *The Industries of Scotland* (1869, David & Charles Reprint 1969), pp 424–9
45 H. Hughes, *The Old Cottages of Snowdonia*, pp 27–8
46 I. C. Peate, *The Welsh House*, Y Cymmrodor, Vol XLVII (London 1940), p 194
47 J. Evans, *Beauties of England and Wales*, Vol XVII, pp 322–3
48 F. J. North, *The Slates of Wales*, pp 52–4
49 E. A. Lewis, *Welsh Port Books, 1550–1663*, Cymmrodorion Record Series, Vol XII, pp ix, xiv, xvi, xxvi, xxx, xxxvi, xlvi, 251, 254–5 and 257
50 A. H. Dodd, *The Industrial Revolution in North Wales*, p 120
51 UCNW, Bangor MS 484

Chapter 2: Penrhyn Quarry, 1731–1782 (pages 27–44)

1 UCNW, Penrhyn 1599 and 1600–1 (original and copy)
2 UCNW, Penrhyn 175–6 (original and copy)
3 UCNW, Penrhyn 177
4 UCNW, Penrhyn 1631
5 UCNW, Penrhyn 133
6 UCNW, Penrhyn 615 and John E. Griffith, *Pedigrees of Anglesey and Caernarvonshire Families*, pp 186–7
7 UCNW, Penrhyn 1638
8 UCNW, Penrhyn 1642
9 UCNW, Penrhyn 1969
10 UCNW, Penrhyn 1643, 1645, 1647 and 1967
11 S. Lewis, *A Topographical Dictionary of Wales*, Vol I
12 UCNW, *Porth yr Aur* 27208 (hereafter *PYA*)
13 A. H. Dodd, *A History of Caernarvonshire, 1284–1900*, pp 194–6
14 UCNW, Penrhyn 1967
15 UCNW, Penrhyn 1969
16 UCNW, Penrhyn 1643
17 UCNW, Penrhyn 1670
18 UCNW, Penrhyn 1657
19 UCNW, Penrhyn 1658
20 P. B. Williams, *The Tourist's Guide through the County of Carnarvon*, pp 103–4, D. Pritchard, 'Random versus Tally Slates', pp 103–4; *Quarry Manager's Journal* (henceforward *QMJ*), August 1946

21 PYA 29081 and F. J. North, *The Slates of Wales*, p 96
22 W. J. Loftie, *Views in North Wales*, p 75
23 Randle Holme, *The Academy of Armory or A Storehouse of Armory and Blazon* (Chester 1688), Book III, pp 209 and 96–7
24 UCNW, Penrhyn 1969
25 UCNW, Penrhyn 1968
26 *Caernarvon and Denbigh Herald*, 5 April 1873 (hereafter *CDH*)
27 UCNW, Penrhyn 1672
28 UCNW, Penrhyn 1687
29 UCNW, Penrhyn 1703
30 UCNW, Penrhyn 1795
31 UCNW, Penrhyn 1675
32 UCNW, Penrhyn 1687
33 *DNB* (London 1895), catalogue of Penrhyn Papers UCNW (1939), pp vi–vii and *Dictionary of Welsh Biography down to 1940* (London 1959).
34 UCNW, Penrhyn 1978–2031
35 UCNW, Penrhyn 1970

Chapter 3: Penrhyn and Dinorwic, 1782–1831 (pages 45–66)

1 *DNB* (London 1895)
2 *North Wales Gazette*, 24 July 1817 (hereafter *NWG*)
3 UCNW, Penrhyn 1971
4 UCNW, Penrhyn 1972
5 Gweirydd Ellis, *A History of the Slate Quarryman in Caernarvonshire in the Nineteenth Century* (unpublished MA thesis), p 33
6 Parliamentary Papers, Vol 48, 1837–8, Paper 215
7 UCNW, Penrhyn 1252
8 Thomas Pennant, *Tours in Wales*, Vol III, p 86
9 UCNW, Penrhyn 1255
10 UCNW, Penrhyn 619
11 UCNW, Penrhyn 620
12 S. Lewis, *A Topographical Dictionary of Wales*, Vol I
13 J. Evans, *A Tour through Part of North Wales in the Year 1798, and at Other Times*, pp 232–3 and UCNW, Penrhyn 1971
14 Thomas Pennant, *Tours in Wales*, Vol III, p 87
15 Thomas Pennant, *Tours in Wales*, Vol II, p 323 and R. T. Pritchard, 'The Post Road in Caernarvonshire', p 21, *TCHS*, Vol 13, 1952
16 P. B. Williams, *The Tourist's Guide through the County of Carnarvon*, p 39

17 E. Pugh, *Cambria Depicta*, pp 99–100 and C. E. Lee, *The Penrhyn Railway*, p 6
18 *NWG*, 26 January 1808
19 P. B. Williams, *The Tourist's Guide through the County of Carnarvon*, p 31 and G. Ellis, *A History of the Slate Quarryman in Caernarvonshire in the Nineteenth Century*, p 41
20 Walter Davies, *General View of the Agriculture of North Wales*, p 84
21 W. Bingley, *A Tour Round North Wales performed during the Summer of 1798*, Vol I, p 133
22 P. B. Williams, *The Tourist's Guide through the County of Carnarvon*, pp 35–8
23 Douglas B. Hague, 'Penrhyn Castle', *TCHS*, Vol 20, 1959, pp 36 and 43
24 UCNW, Penrhyn 631
25 UCNW, Penrhyn 640
26 UCNW, Penrhyn 2039–50
27 Parliamentary Papers 1823, XV (445), pp 351–4
28 UCNW, Penrhyn 711
29 UCNW, Penrhyn 1975
30 UCNW, Penrhyn 2039–50
31 Caernarvonshire Record Office (hereafter CRO) Vaynol 245 and PYA 29077
32 PYA 29098
33 PYA 29094
34 PYA 29095
35 PYA 29091
36 PYA 29089
37 PYA 28479
38 PYA 29076
39 PYA 29083
40 PYA 29080
41 PYA 29223
42 PYA 29082 and *CDH*, 10 May 1873
43 PYA 29240
44 PYA 29241
45 PYA 30173
46 PYA 29081
47 PYA 29079, 29194 and 29082
48 PYA 29188
49 J. E. Griffith, *Pedigrees of Anglesey and Caernarvonshire Families*, p 368
50 PYA 29270
51 PYA 29271

52 PYA 29272
53 *NWG*, 26 July 1810
54 *NWG*, 20 July, 21, 28 September and 2 November 1809 and A. H. Dodd, *The Industrial Revolution in North Wales*, pp 77–8
55 UCNW, Bangor 8702, Part II (this is a Xerox copy of MS 3.479, Cardiff Public Library)

Chapter 4: Nantlle and Ffestiniog, 1782–1831 (pages 67–87)

1 PYA 27203
2 PYA 27205
3 PYA 27208
4 PYA 27209 and 27211
5 PYA 27552
6 PYA 27560
7 PYA 27582
8 PYA 27222
9 PYA 27220
10 PYA 27562
11 PYA 27226
12 PYA 27566
13 PYA 27238
14 PYA 27251
15 PYA 27253
16 PYA 27265
17 PYA 27427
18 PYA 27430 and 28672
19 A. H. Dodd, *The Industrial Revolution in North Wales*, p 76
20 PYA 27273
21 PYA 27276
22 PYA 27277
23 PYA 27375
24 PYA 27523
25 PYA 27344
26 PYA 27096
27 PYA 27564 and 29480
28 PYA 27337
29 PYA 29831, 29841, 29844, 29849, 29850, 29855 and 29857
30 PYA 30011
31 PYA 30163a
32 PYA 30192
33 PYA 27436, 29731, 29727, 29787 and 29829

34 PYA 29736
35 PYA 30102
36 PYA 30104
37 PYA 27438, 27387, 27393, 27797 and 29889
38 J. G. Jones, *The Social and Historical Geography of the Ffestiniog Slate Industry* (unpublished MA thesis), p 38
39 E. Pugh, *Cambria Depicta*, pp 167–8
40 National Library of Wales (hereafter NLW), Samuel Holland Papers 4983 C
41 F. J. North, *The Slates of Wales*, pp 43–4
42 J. G. Jones, *The Social and Historical Geography of the Ffestiniog Slate Industry*, p 41; G. J. Williams, *Hanes Plwyf Ffestiniog*, p 82 and NLW, Samuel Holland Papers 4983 C
43 J. E. Vincent (ed), *The Memories of Sir Llewellyn Turner* (London 1903), pp 13, 17 and 38
44 PYA 30364
45 PYA 30403
46 UCNW, Bangor 8702, Part II
47 *CDH*, 12 November 1853 and J. E. Vincent (ed), *The Memories of Sir Llewellyn Turner*, pp 56 and 79
48 NLW, Samuel Holland Papers 4983 C. These memoirs are printed and edited by W. L. Davies, 'The Memoirs of Samuel Holland', *The Merioneth Historical and Record Society*, Series No 1 (1952)
49 UCNW, Penrhyn 2034
50 NLW, Samuel Holland Papers 4983 C
51 G. J. Williams, *Hanes Plwyf Ffestiniog*, p 87
52 *NWG*, 14 April 1825
53 *NWG*, 10 November 1825
54 *NWG*, 27 October 1825
55 *NWG*, 13 July 1826
56 NLW, Samuel Holland Papers 4983 C
57 *CDH*, 22 February 1873

Chapter 5: Economic, Social and Technical Developments, 1782–1831 (pages 88–116)

1 Owen Morris, *Portmadoc and its Resources* (Blaenau Ffestiniog 1856), p 52 and W. Davies, *General View of the Agriculture of North Wales*, p 412
2 H. A. Shannon, 'Bricks—A Trade Index, 1785–1849', *Economica*, No 3, August 1934, pp 301–4 and Dylan Pritchard, *The Slate Industry of North Wales; a Study of the Changes in Eco-*

nomic Organisation from 1780 to the Present Day (unpublished MA thesis), pp 21–2

3 D. Pritchard, 'The Expansionist Phase in the History of the Welsh Slate Industry', *TCHS*, Vol 10, 1949, pp 65–7

4 PYA 29240

5 *Journal of the House of Commons*, XLIX, 6 February 1794

6 PYA 30326 and F. J. North, *The Slates of Wales*, p 38

7 PYA 30328

8 PYA 30331

9 PYA 30332

10 PYA 30335

11 PYA 30534

12 PYA 30334

13 J. Owen Jones, *Wales*, Vol I, No 4, August 1894, p 162 and Dafydd Tomos (ed), *Michael Faraday in Wales*, p 93

14 PYA 29078

15 PYA 28022

16 PYA 28007

17 Elias Owen, 'The Penrhyn Slate Quarry', *The Red Dragon*, Vol VII, p 329

18 E. Hyde Hall, *A Description of Caernarvonshire (1809–1811)*, p 117. PYA 29816, 30160 and 29888

19 George Kay, *Agriculture of North Wales*, pp 16 and 20–1

20 A. Aikin, *Journal of a Tour through North Wales and Part of Shropshire*, p 140

21 PYA 28197

22 PYA 28023

23 D. C. Davies, *Slate and Slate Quarrying*, p 118

24 PYA 29812

25 NLW, Samuel Holland Papers 4987 B

26 E. L. Lewis, *The Slate Industry*, p 8; D. Pritchard, *The Slate Industry of North Wales*, p 135 and NLW 8743 E

27 UCNW, Penrhyn Additional 2913

28 UCNW, Bangor 8702, Part II

29 PYA 28090, 28097, 28180

30 *NWG*, 14 September 1826

31 PYA 30104

32 PYA 30127 and 30129

33 PYA 29644

34 NLW, Glynllivon 2115

35 NLW, Glynllivon 2105

36 NLW, Glynllivon 2120

37 PYA 29807

38 PYA 29844

39 P. B. Williams, *The Tourist's Guide through the County of Carnarvon*, p 35 and *NWG*, 3 August 1826
40 *CDH*, 3 May 1873
41 A. Aikin, *Journal of a Tour Through North Wales and Part of Shropshire*, pp 125–6
42 PYA 28710
43 PYA 28831
44 *NWG*, 20 March 1817
45 *NWG*, 20 April 1817
46 PYA 29661
47 PYA 31163 and 31164
48 D. Pritchard, *The Slate Industry of Wales*, p 49
49 PYA 30213, 30214, 30215 and 30219
50 PYA 30300, 30301, 30303, 30305, 30306, 30309, 30311 and 30321
51 PYA 30346–555
52 PYA 28899
53 PYA 33076–200 (The Samuel Papers)
54 W. M. Richards, *A General Survey of The Slate Industry of Caernarvonshire and Merionethshire* (unpublished MA thesis), pp 37 and 105
55 PYA 31044
56 PYA 28830
57 PYA 28897
58 A. H. Dodd, *The Industrial Revolution in North Wales*, pp 316–17
59 *NWG*, 26 July 1810
60 P. B. Williams, *The Tourist's Guide through the County of Carnarvon*, pp 105 and 114
61 D. C. Davies, *Slate and Slate Quarrying*, pp 117–24
62 D. Pritchard, *The Slate Industry of Wales*, p 37; Morgan Richards, *Slate Quarrying and How to make it Profitable*, p 37 and PYA 28016, 28032 and 28020
63 *NWG*, 2 May 1811 and PYA 29486
64 PYA 29483 and 29648–59
65 PYA 28222
66 PYA 28482
67 PYA 28485
68 PYA 28480 and 29081
69 PYA 28297
70 PYA 30841
71 PYA 30842
72 PYA 30844 and 30850
73 UCNW, Bangor 8702, Part II

74 *CDH*, 10 May 1873
75 PYA 30550
76 E. Pugh, *Cambria Depicta*, p 104; J. Evans, *Tour through Part of North Wales in the Year 1798*, p 233 and E. L. Lewis, *The Slate Industry*, p 16
77 PYA 29240
78 *NWG*, 31 August 1809; PYA 29284, 29286 and 29297.
79 P. B. Williams, *The Tourist's Guide through the County of Carnarvon*, p 85
80 J. I. C. Boyd, *The Festiniog Railway*, Vol I, 1800–1889, pp 3–9 and Desmond King-Hele, *Shelley, His Thought and Work* (London 1962), pp 23–4
81 Thomas Love Peacock, *Headlong Hall* (Chapter VII)
82 Henry Hughes, *Immortal Sails*, pp 24–5
83 W. M. Richards, 'Some Aspects of the Industrial Revolution in South East Caernarvonshire', *TCHS*, Vols 4 and 5, 1944 passim; *NWG*, 24 July 1817 and *North Wales Chronicle* (hereafter *NWC*), 25 September 1828
84 PYA 29954, 29957, 29966 and 29970
85 PYA 30193 and *NWC*, 16 October 1828
86 PYA 30364; UCNW, Nantlle 12 and NLW, Glynllivon 2120
87 *NWG*, 27 January and 24 March 1825
88 UCNW, Coetmor Papers 49
89 *CDH* 6, 20 and 27 June 1874
90 PYA 29755
91 *Caban*, February 1952 and *CDH*, 17 May 1873
92 George Nicholson, *The Cambrian Traveller's Guide and Pocket Companion*, p 299
93 W. Davies, *General View of the Agriculture of North Wales*, p 99
94 UCNW, Bangor 8702, Part II
95 PYA 29297c
96 *NWC*, 22 May 1828
97 CRO, Vaynol 2523
98 CRO, Vaynol 2524
99 CRO, Vaynol 2554. See also Mary A. Aris, 'Dinorwic Quarry in 1829: A Friendly Society and a Quarry Shop', CRO *Bulletin*, No 3, 1970
100 PYA 28615
101 CRO, Vaynol 2534
102 PYA 30345a
103 NLW 4983 C

Chapter 6: Expansion in Output, 1831–1882 (pages 117–51)

1 H. A. Shannon, 'Bricks—A Trade Index, 1785–1849', *Economica*, August 1934, p 303
2 CRO, Vaynol 2533
3 Nigel Harvey, *A History of Farm Buildings in England and Wales*, p 198
4 Charles Hadfield, *British Canals* (Newton Abbot 1970), p 214
5 Jean Lindsay, *The Canals of Scotland* (Newton Abbot 1968), p 91
6 Michael Robbins, *The Railway Age*, p 47
7 J. Simmons, *The Railways of Britain* (London 1960), p 10 and Michael Robbins, *The Railway Age*, p 47
8 NLW 4984 C
9 NLW 4983 C
10 W. M. Richards, 'Some Aspects of the Industrial Revolution in South Caernarvonshire', *TCHS*, Vol 5, 1944, pp 73–6 and A. G. W. Garraway, 'The Festiniog Railway', *TCHS*, Vol 16, 1955, p 74
11 J. G. Jones, *The Social and Historical Geography of the Ffestiniog Slate Industry*, p 215
12 CRO, Dorothea Quarry Minute Book No 5
13 D. Pritchard, *The Slate Industry of North Wales*, p 95
14 CRO, Dorothea Quarry Minute Book No 5
15 PYA 29889
16 *CDH*, 12 April 1873
17 UCNW, Bangor 1384
18 PYA 30104, 30127 and 30129; *CDH*, 19 April 1873 and *Slate Trade Gazette*, Vol XI, No 91, August 1905, p 141
19 UCNW, Nantlle Papers 19; D. Pritchard, *The Slate Industry of North Wales*, p 101; Parliamentary Papers 39, LXXXV and *Slate Trade Gazette*, Vol XVI, No 155, December 1910
20 UCNW, Cynhaiarn 166; Elizabeth Beazley and M. J. T. Lewis, 'Gorseddau Slate Quarry', *Architectural Review*, 1964 (CRO M/1634); Charles E. Lee, *The Welsh Highland Railway* (Dawlish 1962), p 6; D. S. Whicher, 'Factories that have withstood Rail and Storm', *Country Quest*, January 1972, p 3; *CDH*, 14 June 1873; D. Pritchard, *The Slate Industry of North Wales*, p 103; D. C. Davies, *Slate and Slate Quarrying*, p 44 and J. I. C. Boyd, *Narrow Gauge Railways in South Caernarvonshire*, pp 17–22. See also pp 25–31
21 *CDH*, 14 June 1873; D. Pritchard, *The Slate Industry of North*

Wales, p 103; D. C. Davies, *Slate and Slate Quarrying* and *Slate Trade Gazette*, Vol XI, No 88, May 1905, p 67

22 *CDH*, 15 March 1873

23 UCNW, Searell 4 and 7

24 NLW, Glynllivon Deed 518

25 UCNW, Baron Hill 6450, 6451, 6458, 6460, 6461, 6462 and 6463

26 *CDH*, 22 March 1873; NLW, Gaewern Quarry (unclassified); Merionethshire Record Office (hereafter MRO) M/614 and J. I. C. Boyd, *Narrow Gauge Railways in Mid-Wales*, pp 19–23

27 *CDH*, 22 March 1873 and MRO M/614

28 D. C. Davies, *Slate and Slate Quarrying*, pp 128–30

29 *CDH*, 22 March 1873 and D. C. Davies, *Slate and Slate Quarrying*, p 64

30 *CDH*, 15 March 1873 and J. B. Snell, 'The Bryneglwys Slate Quarry', *Talyllyn Adventure*, L. T. C. Rolt (ed), (Newton Abbot 1971), pp 181–2

31 *CDH*, 5 April 1873 and R. Christiansen and R. W. Miller, *The Cambrian Railways*, Vol I, 1852–1888 (Newton Abbot 1967), p 49

32 D. C. Davies, *Slate and Slate Quarrying*, pp 65–71

33 NLW, Nant Gwryd Papers 12512–16; MRO M/614; *Slate Trade Gazette*, Vol XI, No 91, 17 August 1905, p 134 and D. C. Davies, *Slate and Slate Quarrying*, pp 70–1

34 J. B. Snell, 'The Bryneglwys Slate Quarry', p 180

35 *CDH*, 8 February 1873 and *Slate Trade Gazette*, Vol XI, No 88, 18 May 1905, p 82

36 Charles E. Lee, *The Welsh Highland Railway*, pp 5–7 and MRO M/746

37 G. J. Williams, *Hanes Plwyf Ffestiniog*, p 83; CRO M/1634 and *CDH*, 22 February 1873

38 *Slate Trade Gazette*, Vol VIII, No 59, 17 December 1902, p 229; UCNW, Cynhaiarn 205 and W. M. Richards, *A General Survey of the Slate Industry of Caernarvonshire and Merionethshire*, p 32

39 G. J. Williams, *Hanes Plwyf Ffestiniog*, pp 93–4; *CDH*, 22 February 1873; J. G. Jones, *The Social and Historical Geography of the Ffestiniog Slate Industry*, p 65 and MRO, V & B DA/G (unclassified).

40 *CDH*, 22 February 1873; G. J. Williams, *Hanes Plwyf Ffestiniog*, p 89 and NLW 4984 C

41 *CDH*, 1 March 1873 and G. J. Williams, *Hanes Plwyf Ffestiniog*, p 90

42 *CDH*, 1 March 1873; Anon, *The North Wales Slate Quarries*

(booklet, CRO 1920) and J. G. Jones, *The Social and Historical Geography of the Ffestiniog Slate Industry*, p 63

43 *CDH*, 1 March 1873

44 *CDH*, 1 March 1873 and W. M. Richards, *A General Survey of the Slate Industry of Caernarvonshire and Merionethshire*, p 30

45 H. C. Casserley, *Railway History in Pictures, Wales and the Welsh Border Counties* (Newton Abbot 1970), p 101; *CDH*, 10 May 1873; Mary A. Aris, 'The Dinorwic Quarrymen's Train Service'; CRO *Bulletin*, No 4, 1971, p 19; *NWC*, 17 March 1860; UCNW, Bangor 5474, No 51 and H. R. Davies, *The Conway and the Menai Ferries*, p 317

46 *CDH*, 17 May 1873; J. E. Griffith, *Pedigrees of Anglesey and Caernarvonshire Families*, p 368; CRO M/1236 and D. Morgan Rees, 'North Wales Quarrying Museum Dinorwic' (leaflet). See also 'The Water Wheel at Dinorwic Quarries (at work 1870 to 1925)', CRO *Bulletin* No 5, 1972, pp 42–3

47 UCNW, Bangor 419 and 420

48 NLW, Glynllivon 2124 (printed circular)

49 UCNW, Cefn Du 2, 4 and PYA 27096

50 UCNW, Cefn Du 19, 75, 76 and Parliamentary Accounts and Papers 39, LXXXV

51 *CDH*, 28 June 1873

52 CRO M/495; *CDH*, 23 June 1873; Parliamentary Papers 39, LXXXV and see also J. Lindsay, 'The Scheme for a Railway between Aber and Bryn Hafod y Wern 1846–1854', CRO *Bulletin* No 4, 1971, pp 23–4

53 J. E. Griffith, *Pedigrees of Anglesey and Caernarvonshire Families*, p 187; *Dictionary of Welsh Biography*; UCNW, Penrhyn 2056–2065; G. Ellis, *A History of the Slate Quarryman in Caernarvonshire*, p 44; Elias Owen, 'The Penrhyn Slate Quarry', *The Red Dragon*, Vol VII, p 332; F. J. North, *The Slates of Wales*, p 93; Parliamentary Papers 39, LXXXV and D. C. Davies, *Slate and Slate Quarrying*, p 171

Chapter 7: The Mechanisation of the Quarries (pages 152–69)

1 D. C. Davies, *Slate and Slate Quarrying*, pp 110–11

2 UCNW, Searell No 9

3 UCNW, Searell No 13

4 PYA 28032

5 *CDH*, 13 May 1848

6 *CDH*, 12 September 1874

7 *NWC*, 16 October 1828

8 J. G. Jones, *The Social and Historical Geography of the Ffestiniog Slate Industry*, p 68
9 D. C. Davies, *Slate and Slate Quarrying*, pp 112–13
10 PYA 29480 and 29486 (the latter MS states that the steam engine was erected 'about July 1812', but this must be inaccurate)
11 PYA 27766 and 28020
12 *CDH*, 22 February 1873; J. G. Jones, *The Social and Historical Geography of the Ffestiniog Slate Industry*, p 69 and O. Morris, *Portmadoc and its Resources*, p 47
13 *CDH*, 1 March 1873
14 *CDH*, 15 March 1873
15 CRO, Dorothea No 5
16 UCNW, Cefn Du 19 and 741
17 J. G. Jones, *The Social and Historical Geography of the Ffestiniog Slate Industry*, p 69 and *QMJ*, Vol 48, No 2, February 1964, p 69
18 *CDH*, 1 March 1873 and Michael Burn, *The Age of Slate*, p 6
19 *CDH*, 12 April 1873
20 O. Morris, *Portmadoc and its Resources*, p 47
21 *Slate Trade Gazette*, Vol III, No 7, October 1896, p 18
22 CRO, Printed Brochure Oakeley Slate Quarries Company Ltd (1755–1937) and *Caban*, January 1950, p 11
23 *Slate Trade Gazette*, Vol III, No 7, October 1896, pp 15–18; CRO, *The Commercial Newspaper*, 27 June 1929 and information supplied by Mr R. H. Davies, manager of Llechwedd Slate Quarry
24 CRO, Dorothea No 5
25 CRO, *The Commercial Newspaper*, 27 June 1929 and *QMJ*, Vol 48, February 1964, p 68
26 PYA 29523
27 UCNW, Penrhyn Additional 2913–15
28 *Slate Trade Gazette*, Vol III, No 7, October 1893, pp 15–18 and Vol VIII, No 58, November 1902, p 205
29 MRO DA/G (unclassified)
30 UCNW, Bangor 5276, pp 23–44; *Slate Trade Gazette*, Vol XVII, No 157, 16 February 1911, p 9 and D. C. Davies, *Slate and Slate Quarrying*, pp 119–20
31 *Slate Trade Gazette*, Vol XI, No 95, 20 December 1905, p 233
32 *Slate Trade Gazette*, Vol XII, No 101, 16 June 1906, p 148
33 UCNW, Rhiwbach Papers No 37
34 *Minutes of Evidence before the Royal Commission on Metalliferous Mines and Quarries*, Vol III, Cd 7478, 1914, Quest 2088
35 D. C. Davies, *Slate and Slate Quarrying*, p 126; E. L. Lewis, *The Slate Industry*, p 13 and CRO, Dorothea No 5
36 D. C. Davies, *Slate and Slate Quarrying*, pp 126–7

37 J. G. Jones, *The Social and Historical Geography of the Ffestiniog Slate Industry*, p 92
38 CRO, Dorothea 1240
39 CRO, PQ 99/4 and 5; *Slate Trade Gazette*, Vol II, January 1896, pp 48 and 78
40 *Slate Trade Gazette*, Vol VIII, No 53, June 1902, p 89
41 CRO M/813/16.
42 J. G. Jones, *The Social and Historical Geography of the Ffestiniog Slate Industry*, p 99 and *Royal Commission on Labour*, Vol I, 1892, Quest 7006
43 *Penrhyn Slates* (brochure), I am indebted to Mr H. Crosby, Sales Manager, Penrhyn Quarries Ltd, for this information
44 *QMJ*, Vol 48, No 2, February 1964, pp 67–70 and information supplied by Mr W. H. Smith, Manager of Penrhyn Quarries Ltd

Chapter 8: Improved Transport and Its Consequences (pages 170–84)

1 D. Pritchard, 'Aspects of the Slate Industry', *QMJ*, April 1944
2 NLW 4983 C
3 PYA 29746
4 Most of this section is based on A. H. Dodd, 'The Roads of North Wales 1750–1850', *Archaeologia Cambrensis*, Vol 80, Seventh Series, Vol V, pp 122–48 and NLW 4983 C
5 W. M. Richards, *A General Survey of the Slate Industry of Caernarvonshire and Merionethshire*, p 101 and C. E. Lee, *The Penrhyn Railway*, pp 11–12
6 UCNW, Penrhyn 1800 and C. E. Lee, *The Penrhyn Railway*, p 29
7 Richard Bagnold Jones, *British Narrow Gauge Railways* (London 1958), pp 66–9
8 Ibid, pp 48–57
9 *Report of the Departmental Committee upon Merionethshire Slate Mines*, 1895, Quests 7692, 2828, 2830 and 79
10 Yearly accounts of the Festiniog Railway Company, Portmadoc (quoted in J. Gordon Jones, *The Social and Historical Geography of the Ffestiniog Slate Industry*
11 J. I. C. Boyd, *The Festiniog Railway*, Vol I, 1800–1899, pp 30–1
12 NLW, Glynllivon 2248
13 J. I. C. Boyd, *The Festiniog Railway*, Vol I, 1800–1889, p 27
14 Charles E. Lee, *The Welsh Highland Railway*, passim
15 R. Christiansen and R. W. Miller, *The Cambrian Railways*, Vol I, 1852–1888, pp 69–70

16 J. G. Jones, *The Social and Historical Geography of the Ffestiniog Slate Industry*, p 166
17 R. B. Jones, *British Narrow Gauge Railways*, pp 56–7
18 J. I. C. Boyd, *Narrow Gauge Railways in Mid-Wales*, pp 19–26 and R. B. Jones, *British Narrow Gauge Railways*, pp 27–8
19 L. T. C. Rolt, *Railway Adventure*, pp 12–13 and 23–5; R. B. Jones, *British Narrow Gauge Railways*, pp 29–31 and *Slate Trade Gazette*, Vol 13, No 118, November 1907, p 212
20 J. I. C. Boyd, *Narrow Gauge Railways in Mid-Wales*, p 125 and R. B. Jones, *British Narrow Gauge Railways*, pp 40–2
21 R. Christiansen and R. W. Miller, *The Cambrian Railways*, Vol II, 1889–1968, pp 42–5
22 J. B. Snell, 'The Bryneglwys Slate Quarry', *Talyllyn Adventure, Railway Adventure and Talyllyn Century* (combined edition, Newton Abbot 1971), p 180
23 J. I. C. Boyd, *Narrow Gauge Railways in Mid-Wales*, p 290
24 W. M. Richards, *A General Survey of the Slate Industry of Caernarvonshire and Merionethshire*, pp 156–8
25 CRO, PQ 90/2, 1920 and *Slate Trade Gazette*, Vol IV, No 242, December 1897, p 47
26 CRO, PQ 90/2, 1920
27 CRO, PQ 90/2, 1917
28 CRO, PQ 90/2, 1919, 1920 and *Slate Trade Gazette*, Vol 21, No 205, February 1915, p 3
29 CRO, PQ 90/2, 1933
30 CRO, PQ 90/2, 1922
31 CRO, PQ 90/2, 1926 and 1937
32 Furnished by Harbour Office, Caernarvon
33 CRO, DQ 2631–53

Chapter 9: The Expansion of Home and Overseas Markets (pages 185–98)

1 D. Pritchard, 'The Expansionist Phase in the History of the Welsh Slate Industry', *TCHS*, Vol 10, 1949, p 67
2 NLW, Llwynbedr 6, 4 April 1842
3 UCNW, Bangor 1384
4 D. Pritchard, 'The Expansionist Phase in the History of the Welsh Slate Industry', *TCHS*, Vol 10, 1949, pp 67–8
5 UCNW, Penrhyn 2034
6 PYA 30271
7 PYA 30225
8 PYA 30550–555

9 UCNW, Penrhyn 1800
10 *CDH*, 26 February 1842
11 UCNW, Penrhyn 1800
12 J. Gordon Jones, 'The Ffestiniog Slate Industry: The Industrial Pattern, 1831–1913, Part I, 1831–1877, A Period of Expansion—The General Background', *Journal of the Merionethshire Historical and Record Society*, Vol VI, Part II, 1970, p 192
13 O. Morris, *Portmadoc and its Resources*, pp 49–51 and Henry Hughes, *Immortal Sails*, p 59
14 W. H. B. Court, *A Concise Economic History of Britain* (Cambridge 1964), pp 189–90
15 O. Morris, *Portmadoc and its Resources*, p 51
16 J. Gordon Jones, 'The Ffestiniog Slate Industry: The Industrial Pattern, 1831–1913, Part I, 1831–1877, A Period of Expansion—The General Background', *Journal of the Merionethshire Historical and Record Society*, Vol VI, Part II, 1970, p 192
17 MRO M/1720
18 UCNW, Cefn Du 737
19 D. Pritchard, *The Slate Industry of North Wales*, p 87
20 W. H. B. Court, *A Concise Economic History of Britain*, p 195
21 D. Pritchard, 'Aspects of the Slate Industry', *QMJ*, September 1943, p 174

Chapter 10: The Growth of Unionism (pages 199–218)

1 *CDH*, 15 February 1873
2 *CDH*, 12 July 1852
3 Quoted in J. Owain Jones, *The History of the Caernarvonshire Constabulary, 1856–1950*, p 11
4 *CDH*, 5 December 1846
5 *CDH*, 29 June 1850
6 G. J. Williams, *Hanes Plwyf Ffestiniog*, p 112
7 D. Pritchard, *The Slate Industry of North Wales*, p 139
8 UCNW, Searell 3
9 NLW 10348 A
10 George Howell, *Labour Legislation, Labour Movements, Labour Leaders* (London 1902), pp 127–45
11 J. Roose Williams, 'Quarryman's Champion. The Life and activities of William John Parry of Coetmor', *TCHS*, Vol 23, 1962, pp 96–107 and Vol 24, 1963, pp 220–1
12 J. Roose Williams, 'Quarryman's Champion', *TCHS*, Vol 24, 1963, pp 222–3
13 NLW 4984 C

14 *Royal Commission on Labour*, Vol 2, 1892, Quests 9541 and 9543
15 *CDH* 7, 14 and 28 November 1868
16 NLW 8753 E
17 *CDH*, 13 December 1873
18 *CDH*, 20 December 1873
19 J. E. Griffith, *Pedigrees of Anglesey and Caernarvonshire Families*, p 368
20 *CDH*, 6 June 1874
21 J. Roose Williams, 'Quarryman's Champion', *TCHS*, Vol 24, 1963, p 227
22 UCNW, Coetmor 49
23 *Royal Commission on Labour*, Vol 2, Quest 9548
24 W. J. Parry, *The Penrhyn Lock-out 1900–1901* (London 1901), p 8
25 *CDH*, 6 June 1874
26 *CDH*, 20 June 1874
27 *CDH*, 18 July 1874
28 UCNW, Coetmor 43, p 8
39 *Royal Commission on Labour*, Vol 2, Quests 9555 and 9556
30 NLW 8752 E
31 W. J. Parry, *The Penrhyn Lock-out 1900–1901*, pp 13 and 28–9
32 *CDH*, 19 September 1874
33 W. J. Parry, *The Penrhyn Lock-out 1900–1901*, pp 32–4
34 W. H. B. Court, *A Concise Economic History of Britain*, pp 248–9
35 W. J. Parry, *The Penrhyn Lock-out 1900–1901*, p 35

Chapter 11: Social Conditions (pages 219–45)

1 UCNW, Searell 3 and 10
2 MRO M/1720
3 UCNW, Bangor 1384
4 CRO PQ 99/5
5 CRO, Dorothea 672
6 *Royal Commission on Labour*, Vol 2, 1892, Quest 9069
7 J. Owen Jones, *Wales*, Vol I, No 4, August 1894, p 166
8 *Slate Trade Gazette*, Vol 20, No 202, November 1914, p 216
9 Ibid, Vol 17, No 163, September 1911, p 159
10 Ibid, Vol 20, No 196, May 1914, p 76
11 CRO DQ 1720
12 J. Owen Jones, *Wales*, Vol I, No 4, August 1894, p 166
13 CRO M/1720
14 CRO M/813/16
15 W. J. Parry, *Hanes Fy Mywyd a'm Gwaith* (nd), p 83

16 CRO M/1573. I am indebted to Mr Bryn Parry for assistance in the interpretation of these letters
17 NLW 4984 C
18 J. Roose Williams, 'Quarryman's Champion', *TCHS*, Vol 25, 1964, p 91
19 *Slate Trade Gazette*, Vol 12, No 101, June 1906, pp 137 and 140
20 Ibid, Vol 14, No 128, September 1908, p 425
21 J. Gordon Jones, *The Social and Historical Geography of the Ffestiniog Slate Industry*, p 224
22 Census of 1841 and G. Ellis, *A History of The Slate Quarryman in Caernarvonshire in the Nineteenth Century*, p 207
23 Census of 1851 and 1861, and J. Gordon Jones, *The Social and Historical Geography of the Ffestiniog Slate Industry*, p 214
24 Census Returns and D. Pritchard, *The Slate Industry of North Wales*, p 124
25 Census Returns
26 A. H. Dodd, *A History of Caernarvonshire, 1284–1900*, p 253
27 UCNW, Carter Vincent 2686 and W. J. Parry, *The Cry of the People*, pp 149–50
28 *Royal Commission on Land in Wales and Monmouthshire*, Vol I, Quests 11,941 and 11,942
29 *CDH*, 1 January 1870
30 Census Returns
31 *Royal Commission on Land in Wales and Monmouthshire*, Vol I, Quests 13,381 and 13,536 and J. Eardley Wilmot, *Reminiscences of the Late Thomas Assheton Smith Esq.* (London 1862), p 76
32 *Royal Commission on Land in Wales and Monmouthshire*, Vol I, Quests 12,073, 12,076–7, 12,080 and 12,083 and *Royal Commission on Labour*, Vol II, 1892, Quest 16,844
33 *Royal Commission on Land in Wales and Monmouthshire*, Vol I, Quests 13,792 and 13,888
34 Ibid, Vol I, Quests 12,069 and 13,488
35 CRO, Vaynol 2968
36 CRO, Vaynol 2972
37 *Royal Commission on Land in Wales and Monmouthshire*, Vol I, Quests 13,400 and 11,967
38 UCNW, Bangor 5474, No 60
39 UCNW, Bangor 5474, No 51
40 UCNW, Bangor 5474, No 56
41 Census Returns
42 Ibid
43 D. Pritchard, *The Slate Industry of North Wales*, p 105
44 *Royal Commission on Labour*, Vol III, Quest 24,641 and UCNW, Rhiwbach 9

45 *Report of the Departmental Committee upon Merionethshire Slate Mines* (1895), XXIX, p 37 and UCNW, Bangor 1521
46 NLW 4984 C
47 *Minutes of Evidence of the Departmental Committee upon Merionethshire Slate Mines*, Quest 1,648
48 Census Returns
49 PYA 27296
50 *CDH*, 14 November 1874
51 G. J. Williams, *Hanes Plwyf Ffestiniog*, p 136
52 J. Gordon Jones, *The Social and Historical Geography of the Ffestiniog Slate Industry*, p 229
53 *Minutes of Evidence of the Departmental Committee upon Merionethshire Slate Mines*, Quest 4, 198
54 *Herald Gymraeg*, 4 December 1865
55 *Report upon Merionethshire Slate Mines*, XXIX, p 40
56 *Slate Trade Gazette*, Vol VIII, No 58, November 1902, p 203
57 CRO, DQ 2269 and 1720
58 *The Welsh Slate Industry, Report by the Committee appointed by the Minister of Works. A Report of an Investigation into the Alleged High Mortality Rate from Tuberculosis of the Respiratory System among Slate Workers in the Gwyrfai Rural District* (HMSO 1927); *Report on an Inquiry into the Occurrence of Disease of the Lungs from Dust Inhalation in the Slate Industry in the Gwyrfai District* (HMSO 1930); *Report of an Investigation into the Causes of the continued High Death-rate from Tuberculosis in certain parts of North Wales* (Cardiff 1933); T. W. Wade, *The Part Played in the Production of Tuberculosis by Environmental Conditions* (Cardiff 1933) and T. W. Davies, 'Silicosis in Slate Quarry Miners', *Journal of Disease of the Chest*, Vol XX, September 1939, p 543
59 J. G. Jones, *The Social and Historical Geography of the Ffestiniog Slate Industry*, p 96 and *QMJ*, Vol 48, No 2, February 1964, p 70
60 *Report upon Merionethshire Slate Mines*, XXIX, pp 81 and 37
61 *Royal Commission on Labour*, Vol II, Quest 16,790
62 *Report upon Merionethshire Slate Mines*, XXIX, pp 38 and 78
63 *Minutes of Evidence taken before the Royal Commission on Metalliferous Mines and Quarries*, Vol I (1912), Quests 651–2
64 *Minutes of Evidence taken before the Royal Commission on Metalliferous Mines and Quarries*, Vol II (1914), Quest 277 and Michael Burn, *The Age of Slate*, pp 13–14
65 MRO M/719
66 NLW 4984 C
67 MRO M/719

68 *Minutes of Evidence of the Departmental Committee upon Merionethshire Slate Mines*, Quests 3,545 and 3,988
69 *Minutes of Evidence taken before the Royal Commission on Metalliferous Mines and Quarries*, Vol I (1912), Quest 579 and CRO PQ 99/4
70 CRO DQ 1720
71 UCNW, Bangor 460 and *Slate Trade Gazette*, Vol 16, No 16, December 1910, p 251
72 J. G. Jones, *The Social and Historical Geography of the Ffestiniog Slate Industry*, p 238
73 A. H. Dodd, *A History of Caernarvonshire, 1284–1900*, p 301
74 NLW 8743 E
75 CRO PQ 100/77
76 CRO PQ 99/7
77 A. H. Dodd, *A History of Caernarvonshire 1284–1900*, p 378
78 *CDH*, 19 March 1842
79 UCNW, Penrhyn 1800
80 A. H. Dodd, *A History of Caernarvonshire 1284–1900*, p 378
81 *Royal Commission on Labour*, Vol II (1892), Quest 16,868
82 UCNW, Cefn Du 740
83 *Royal Commission on Labour*, Vol III (1893), Quests 24,720–3 and 24,827–8
84 *Royal Commission on Labour*, Vol II (1892), Quest 16,889
85 CRO, Penyrorsedd Rules 1899 (unclassified)
86 *CDH*, 31 January 1846
87 CRO PQ 99/1

Chapter 12: Depression and Decline, 1882–1918 (pages 246–60)

1 W. H. B. Court, *A Concise Economic History of Britain*, p 195
2 W. M. Richards, *A General Survey of the Slate Industry of Caernarvonshire and Merionethshire*, p 46
3 D. Pritchard, 'Aspects of the Slate Industry', *QMJ*, May 1946, p 559
4 *Report on Slates prepared by a Sub-Committee appointed by the Standing Committees on the Investigation of Prices and Trusts*, 1921, Cmd 1338
5 UCNW, Cefn Du 737
6 UCNW, Cefn Du 739
7 UCNW, Cefn Du 740
8 CRO PQ 26/1
9 CRO PQ 99/1
10 D. Pritchard, *The Slate Industry of North Wales*, p 176

11 F. J. North, *The Slates of Wales*, p 103
12 Tariff Figures (quoted in D. Pritchard, *The Slate Industry of North Wales*, p 176)
13 CRO M/813/16
14 *Slate Trade Gazette*, Vol 2, No 3, December 1895, pp 28–31
15 Ibid, Vol 2, No 4, January 1896, p 96
16 MRO DA/G (unclassified)
17 CRO PQ 99/3
18 CRO PQ 99/5
19 *Slate Trade Gazette*, Vol 2, No 3, December 1895, pp 53–8 and CRO 99/5
20 UCNW, Cefn Du 747
21 CRO M/1573
22 *Slate Trade Gazette*, Vol 5, No 25, December 1898, pp 56–7 and 81
23 CRO PQ 99/5
24 CRO PQ 99/5
25 CRO M/813/16
26 CRO, Dorothea No 5
27 *Slate Trade Gazette*, Vol 9, No 69, October 1903, p 235
28 Ibid, Vol 9, No 94, November 1905, p 215
29 Ibid, Vol 9, No 93, October 1905, No 94, November 1905 and NLW 8759 E
30 *Slate Trade Gazette*, Vol 12, No 97, February 1906, p 12
31 Ibid, Vol 12, No 98, March 1906, pp 31 and 35 and No 103, August 1906, p 140
32 Ibid, Vol 12, No 104, September 1906, pp 143 and 149 and Vol 13, No 111, April 1907, p 44
33 Ibid, Vol 13, No 118, November 1907, p 216
34 Ibid, Vol 14, No 124, May 1908, p 340
35 Ibid, Vol 14, No 128, September 1908, p 425 and No 131, December 1908, p 494
36 Ibid, Vol 14, No 143, December 1909, p 239
37 *Collected Works of William Morris* (London 1914), Vol 22, p 407
38 *Slate Trade Gazette*, Vol 15, No 152, September 1910, p 179
39 Ibid, Vol 20, No 200, September 1914 and CRO JWG 52
40 *Slate Trade Gazette*, Vol 20, No 201, November 1914, p 206 and Vol 21, No 207, April 1915, p 39
41 *Report on Slates prepared by a sub-committee appointed by the Standing Committee on the Investigation of Prices and Trusts*, 1921, Cmd 1338, p 4

Chapter 13: Industrial Unrest (pages 261–83)

1 NLW 8736 C
2 D. Pritchard, *The Slate Industry of North Wales*, pp 203–5
3 *Royal Commission on Labour*, Vol III (1893), Quests 9570–1, 9581–2 and 9745
4 Anon, *The Lock-out at the Dinorwic Quarries 1885* (nd) and UCNW, Cefn Du 740
5 *Y Genedl*, 5 March 1886
6 MRO DA/G Minute Book 1874–1901 (unclassified)
7 *NWC*, 10 and 17 April 1880
8 W. J. Parry, *The Penrhyn Lock-out 1900–1901*, pp 35–51
9 CRO PQ 99/1
10 CRO PQ 99/3 and 4
11 CRO PQ 99/3 and 4
12 CRO, Dorothea No 5
13 W. J. Parry, *The Penrhyn Lock-out 1900–1901*, p 71
14 NLW 8752 E
15 CRO PQ 99/4 and 5
16 W. J. Parry, *The Penrhyn Lock-out 1900–1901*, pp 71–90
17 W. J. Parry, *The Penrhyn Lock-out 1900–1901*, pp 103–6
18 CRO PQ 99/5
19 W. J. Parry, *The Penrhyn Lock-out 1900–1901*, pp 164–8
20 CRO PQ 99/5
21 CRO PQ 99/5
22 W. J. Parry, *The Penrhyn Lock-out 1900–1901*, p 169
23 NLW 8752 E
24 W. J. Parry, *The Penrhyn Lock-out 1900–1901*, p 171
25 NLW 8746 E
26 W. J. Parry, *The Penrhyn Lock-out 1900–1901*, pp 174–5 and *NWC*, 14 March 1903
27 *Slate Trade Gazette*, Vol 7, No 43, August 1901, p 181 and No 44 September 1901, pp 185, 213–14 and 233
28 Ibid, Vol 7, No 47, December 1901, p 276
29 Ibid, Vol 7, No 48, January 1902, p 303
30 CRO SJC Copy of Evidence of a Sub-Committee of the Standing Joint (Police) Committee, December 1902, Quests 62, 108, 700, 719, 721 and 828
31 The North Wales Quarrymen's Union, *The Penrhyn Quarry Dispute 1900–1903* (1904), p vi
32 C. Sheridan Jones, *What I Saw at Bethesda* (nd), passim
33 NLW 8759 E

34 NLW 8752 E and *Slate Trade Gazette*, Vol 9, No 61, February 1903, p 38 and No 67, August 1903, p 151
35 The North Wales Quarrymen's Union, *The Penrhyn Dispute 1900–1903*, p viii
36 *Slate Trade Gazette*, Vol 9, No 71, December 1903, p 214
37 Ibid, Vol 17, No 163, September 1911
38 J. Roose Williams, 'Quarryman's Champion', *TCHS*, Vol 26, 1965, pp 157–61

Chapter 14: The Changing Structure of the Industry, 1918–1945 (pages 284–97)

1 *Ministry of Labour Gazette* (quoted in D. Pritchard, *The Slate Industry of North Wales*, p 172)
2 CRO M/813/16, p 34
3 W. M. Richards, *A General Survey of the Slate Industry of Caernarvonshire and Merionethshire*, p 46
4 CRO PQ 99/9
5 CRO JWG 57 and CRO DQ 1720
6 CRO, Dorothea No 4
7 *Report on Slates* (1921), Cmd 1338, p 6; W. M. Richards, *A General Survey of the Slate Industry of Caernarvonshire and Merionethshire*, p 210 and CRO DQ 1720
8 CRO, Dorothea 1469
9 CRO, Dorothea 564
10 MRO DA/G Foreign and Colonial Orders Book 1922–1935 (unclassified)
11 D. Pritchard, *The Slate Industry of North Wales*, p 188
12 CRO, Dorothea No 4
13 MRO DA/G (unclassified) and J. G. Jones, *The Social and Historical Geography of the Ffestiniog Slate Industry*, p 75
14 CRO JWG 73 and 69
15 *Report on Slates*, 1921, p 5
16 CRO, Dorothea No 3
17 MRO DA/G (unclassified)
18 *The Welsh Slate Industry*, *Report by the Committee appointed by the Minister of Works*, 10 and D. Pritchard, 'Aspects of the Slate Industry', *QMJ*, May 1946, p 561
19 CRO JWG 73–5
20 N. Harvey, *A History of Farm Buildings in England and Wales*, p 197
21 J. G. Jones, *The Social and Historical Geography of the Ffestiniog Slate Industry*, p 75

22 CRO M/813/15, pp 1–4
23 CRO JWG 73 and *Report of the Standing Committee on Roofing Slates* (1930), Cmd 3461, p 6
24 D. Pritchard, *The Slate Industry of North Wales*, pp 181–2
25 CRO JWG 85/2
26 CRO JWG 73
27 CRO PQ 99/4
28 CRO M/813/16, CRO DQ 1720 and D. Pritchard, *The Slate Industry of North Wales*, p 195
29 D. Pritchard, *The Slate Industry of North Wales, Statement of the Case for a Plan* (1946), p 9
30 *The Welsh Slate Industry*, 25, 14 and 15

Chapter 15: The Slate Industry since 1945 (pages 298–307)

1 *The Welsh Slate Industry*, 7, 8 and Appendix I
2 CRO DQ 1788, and information received from Mr R. H. Davies and Mr W. H. Smith
3 *The Welsh Slate Industry*, 31
4 UCNW, Bangor 11511 and E. B. Kenyon, *The Slate Industry and its Place in the Industrial Structure of North-west Wales* (1951), p 36 (typescript dissertation)
5 CRO, Dorothea No 20
6 CRO JWG 102–6
7 *Annual Statement of the Trade of the United Kingdom*, Vol II (1955), p 296
8 *Annual Statement of the Trade of the United Kingdom*, Vol III (1955), p 392
9 *Annual Statement of the Trade of the United Kingdom*, Vol II (1960), p 248
10 *Digest of Welsh Statistics* (HMSO)
11 J. G. Jones, *The Historical and Social Geography of the Ffestiniog Slate Industry*, p 94
12 CRO, Dorothea 1469
13 CRO DQ 2631–53
14 *QMJ*, March 1962, pp 117–20
15 CRO, *Dinorwic Slate Data*, Autumn 1962, No 7
16 *QMJ*, August 1962, p 304
17 CRO DQ 1816
18 CRO, Dorothea No 21
19 *NWC*, 30 September 1971, brochures from Penrhyn Quarries Ltd, and information received from Mr H. Crosby and Mr W. H. Smith

20 CRO, Dorothea 1469

21 *The Times*, 11 and 24 April 1972

22 Information received from the managers of the following quarries: Maenofferen Slate Quarry Company Ltd, the Caernarvonshire Crown Slate Quarries Ltd, the Cwt y Bugail Slate Quarries Company Ltd, the Penyrorsedd Slate Quarry Company Ltd and Wincilate Ltd

23 A. G. Bradley, *Highways and Byways in North Wales* (London 1898), p 257

24 E. Pugh, *Cambria Depicta*, pp 103–4

25 CRO JWG 238

BIBLIOGRAPHY

MANUSCRIPTS

Bangor (*The University College Library*)
Bangor
Baron Hill
Carter Vincent
Cefn Du
Coetmor
Cynhaiarn
Nantlle
Penrhyn
Penrhyn Additional
Porth yr Aur
Rhiwbach
Searell

Aberystwyth (*National Library of Wales*)
Adrian Stokes manuscripts relating to slate quarries
A collection by W. J. Parry of manuscripts and papers
A diary of William Thomas, Llanwnda
Glynllivon Papers relating to slate quarries
Nant Gwryd Papers
Reports upon Diphwys Quarry 1865–8
Robertson Williams Documents
Samuel Holland Papers

Caernarvon (Caernarvonshire Record Office)
Bangor Royal Slate Company
Dinorwic Quarry
Dorothea
Eagle Red Slate Company
J. W. Greaves
Penrhyn Quarry
Penyrorsedd Quarry
Vaynol Estate Papers

Dolgellau (Merioneth Record Office)
Braichgoch
Croesor
Llechwedd
Oakeley
Rhiwbach
Votty and Bowydd

NEWSPAPERS AND PERIODICALS
Archaeologia Cambrensis
Caernarvon and Denbigh Herald
Herald Gymraeg
North Wales Chronicle
North Wales Gazette
Quarry Manager's Journal
Slate Trade Gazette
The Merioneth Historical and Record Society
The Transactions of the Society of Cymmrodorion
Transactions of the Caernarvonshire Historical Society
Y Cymmrodor
Y Genedl

PARLIAMENTARY REPORTS, PAPERS AND PUBLICATIONS
Annual Trade Returns
Census Returns
Journal of the House of Commons XLIX 1794

Papers relating to the Penryhn and Bryn Hafod y Wern Quarries:
1819 XVI and XVII
1821 XXI
1822 XXI
1823 XII and XV

Papers relating to the Slate Tax, 1830–1 X

Report of the Departmental Committee upon Merionethshire Slate Mines, 1895

Reports of the Royal Commission on Labour, 1891–4

Reports of the Royal Commission on Metalliferous Mines and Quarries, 1912–1914

Royal Commission on Land in Wales and Monmouthshire, Vol 1 (1894)

The Welsh Slate Industry, Report by the Committee Appointed by the Minister of Works (HMSO 1947)

THESES

Ellis, Gweirydd. *A History of the Slate Quarrymen in Caernarvonshire in the Nineteenth Century* (Bangor 1931)

Jones, J. G. *The Social and Historical Geography of the Ffestiniog Slate Industry* (Aberystwyth 1959)

Pritchard, Dylan. *The Slate Industry of North Wales: a Study of the Changes in Economic Organisation from 1780 to the Present Day* (Bangor 1935)

Richards, W. M. *A General Survey of the Slate Industry of Caernarvonshire and Merionethshire* (Liverpool 1933)

PRINTED SOURCES

Aikin, A. *Journal of a Tour through North Wales and Part of Shropshire* (London 1797)

Bingley, W. *A Tour Round North Wales Performed during the Summer of 1798*, Vol 1 (London 1800)

Boyd, J. I. C. *Narrow Gauge Railways in Mid-Wales* (Lingfield 1970)

——. *Narrow Gauge Railways in South Caernarvonshire* (Lingfield 1972)

——. *The Festiniog Railway*, Vol 1, 1800–1889 (Lingfield 1956)

Burn, M. *The Age of Slate* (Blaenau Ffestiniog nd)

Christiansen, R. and Miller, R. W. *The Cambrian Railways*, Vol I and II (Newton Abbot 1967)

Davies, D. C. *Slate and Slate Quarrying* (London 1878)

Davies, H. R. *The Conway and the Menai Ferries* (Cardiff 1966)

Davies, W. *General View of the Agriculture of North Wales* (London 1810)

Dodd, A. H. *A History of Caernarvonshire 1284–1900* (Caernarvonshire Historical Society 1968)

——. *The Industrial Revolution in North Wales* (Cardiff 1951)

Evans J. *A Tour through Part of North Wales in the Year 1798 and at Other Times* (London 1800)

——. *Beauties of England and Wales*, Vol XVII (London 1812)

——. *Letters Written during a Tour through North Wales in the Year 1798 and at Other Times* (Edinburgh 1804)

Gethin, J. O. *Hanes Penmachno a Dolwyddelen* (Llanrwst 1884)

Griffith, J. *Chwarelau Dyffryn Nantlle a Chymydogaeth Moeltryfan* (Conway 1934)

Griffith, J. E. *Pedigrees of Anglesey and Caernarvonshire Families* (Horncastle 1914)

Hall, E. Hyde. *A Description of Caernarvonshire (1809–1811)*, ed E. Gwynne Jones (Caernarvon 1952)

Harvey, N. *A History of Farm Buildings in England and Wales* (Newton Abbot 1970)

Hucks, J. *A Pedestrian Tour Through North Wales* (London 1795)

Hughes, H. *Immortal Sails* (London 1951)

——. *The Old Cottages of Snowdonia* (Bangor 1908)

Jones, J. O. *The History of the Caernarvonshire Constabulary 1856–1950* (Caernarvonshire Historical Society 1963)

Kay, G. *Agriculture of North Wales* (Edinburgh 1794)

Lee, C. E. *The Penrhyn Railway* (Welsh Highland Light Railway Ltd 1972)

——. *The Welsh Highland Railway* (The Welsh Highland Railway Society and David & Charles 1962)

Leigh's *Guide to Wales and Monmouthshire* (London 1839)

Leland, J. *The Itinerary in Wales of John Leland in or about the Years 1536–9*, ed L. T. Smith (London 1906)

Lewis, E. A. *Welsh Port Books (1550–1603)* (London 1927)

Lewis, E. L. *The Slate Industry* (Colorado 1927)

Lewis, S. *A Topographical Dictionary of Wales*, Vol 1 (London 1833)

Loftie, W. J. *Views in North Wales* (London 1875)

Morris, O. *Portmadoc and its Resources* (Blaenau Ffestiniog 1856)

Nicholson, G. *The Cambrian Traveller's Guide and Pocket Companion* (Stourport 1808)

North, F. J. *The Slates of Wales* (Cardiff 1946)

Owen, G. D. *Elizabethan Wales* (Cardiff 1962)

Parry, Cyril. *The Radical Tradition in Welsh Politics* (Hull 1970)

Parry, W. J. *Chwareli a Chwarelwyr* (Caernarvon 1897)

——. *Hanes Fy Mywyd a'm Gwaith* (nd)

——. *The Cry of the People* (Caernarvon 1906)

Pennant, T. *Tours in Wales*, Vols II and III (London 1810)

Pugh, E. *Cambria Depicta* (London 1816)

Richards, M. *Slate Quarrying and How to Make it Profitable* (nd)

Royal Commission on Ancient Monuments in Wales and Monmouthshire, Vols I, II and III (London 1960)

Salzman, L. F. *Building in England down to 1540* (Oxford 1952)

Smith, B and George, T. Neville. *British Regional Geology. North Wales* (London 1961)

Warner, R. *A Second Walk through Wales in August and September 1798*, Vol II (London 1800)

Williams, G. J. *Hanes Plwyf Ffestiniog* (Wrexham 1882)

Williams, J. L. and Williams, Ifor. *Gwaith Guto'r Glyn* (Cardiff 1939)

Williams, P. B. *The Tourist's Guide through the County of Carnarvon* (Caernarvon 1821)

Williams, R. *The History and Antiquities of the Town of Aberconwy* (Denbigh 1835)

ACKNOWLEDGEMENTS

In writing this book, I have been very heavily dependent on the help and goodwill of Record Office and library staff. I should like to express my gratitude to Mr A. Giles Jones and Miss D. Clark of the University College of North Wales for their cheerful and patient attention. Mr Bryn R. Parry of the Caernarvonshire Record Office gave me much valuable information and advice; and Mrs Mary A. Aris catalogued at high speed the records of the Dinorwic and Dorothea Quarries, which came into the office while I was engaged on this research. I should also like to thank Miss Anne E. Jones and Mrs C. Hughes for their help. My application for access to the Penrhyn records for 1900–3 was turned down by Lady Janet Douglas Pennant. Mr Gareth Williams of the Merioneth Record Office helped me in many ways, not least by allowing me to examine some records before they had been catalogued. I am very grateful to the staff of the UCNW Library, and especially to Mr Derwyn Jones and to the late Mr E. Gwynne Jones. I have also to acknowledge the help I received from the staffs of the National Library of Wales, the Manchester Central Library, and the Bangor Public Library.

I received much helpful information about present conditions in the North Wales slate quarries from the following quarry managers: Mr H. Crosby (Sales Manager) and Mr W. H. Smith of Penrhyn Quarries Ltd; Mr J. F. Lloyd of Wincilate Ltd; the managing director of Maenofferen; Mr O. J. Williams of Penyrorsedd; Mr R. H. Davies of Llechwedd; Mr Price of Cwt y Bugail; Mr W. Riley of the Caernarvonshire Crown Slate Quarries Ltd; and Mr I. Hughes of the Vronlog Green Slate

Quarry. I received from the Carnegie Trust for the Universities of Scotland a grant towards research expenses, and from the Twenty-Seven Foundation a grant towards typing expenses. I am also grateful, for assistance of various kinds, to Dr R. S. Fitton, Mr and Mrs Meshach Roberts, Mr A. J. Taylor, Mr W. M. Tydeman, Mr G. A. Usher and to my publishers. I have to thank Mr Geoffrey Charles and Mr D. Rendell for help with the photographs.

Finally, I must acknowledge the encouragement given by my children, Cora and John, and the help given in a great many ways by my husband.

J.L.

Bangor

INDEX

Page numbers in italics indicate photographs